HAIR SHEEP OF WESTERN AFRICA AND THE AMERICAS

A Winrock International Study

Also in This Series

Future Dimensions of World Food and Population,
edited by Richard G. Woods

HAIR SHEEP OF WESTERN AFRICA AND THE AMERICAS

A GENETIC RESOURCE FOR THE TROPICS

edited by H. A. Fitzhugh
and G. E. Bradford

Routledge
Taylor & Francis Group

LONDON AND NEW YORK

First published 1983 by Westview Press

Published 2018 by CRC Press
Taylor & Francis Group
6000 Broken Sound Parkway NW, Suite 300
Boca Raton, FL 33487-2742

CRC Press is an imprint of the Taylor & Francis Group, an informa business

Visit the Taylor & Francis Web site at
http://www.taylorandfrancis.com

and the CRC Press Web site at
http://www.crcpress.com

Library of Congress Cataloging in Publication Data
Main entry under title:
Hair sheep of western Africa and the Americas.
    (A Winrock International study)
    Includes index.
    1. Hair   sheep—America. 2. Hair   sheep—Africa,   West. 3. Hair   sheep—Tropics.
I. Fitzhugh, H. A., 1939–   . II. Bradford, G. E. (Gordon Eric), 1929–   III.
Winrock International Livestock Research and Training Center.   IV. Series.
SF373.H17H37   1982   636.3'145   82-16032

ISBN 13: 978-0-367-01925-9 (hbk)
ISBN 13: 978-0-367-16912-1 (pbk)

CONTENTS

Preface.. ix
Acknowledgments... xi
The Contributors... xiii

SECTION 1
GENERAL OVERVIEW

1.1 Hair Sheep: A General Description,
G. E. Bradford and H. A. Fitzhugh........................... 3

1.2 Productivity of Hair Sheep and Opportunities
for Improvement, H. A. Fitzhugh
and G. E. Bradford... 23

SECTION 2
MIDDLE AMERICA AND SOUTH AMERICA

2.1 Pelibuey Sheep in Mexico, Mario Valencia Zarazúa
and Everardo Gonzalez Padilla............................. 55

2.2 Reproduction in Peliguey Sheep,
A. Gonzalez-Reyna, J. De Alba,
and W. C. Foote... 75

2.3 African Sheep in Colombia, Rodrigo Pastrana B.,
Rafael Camacho D., and G. E. Bradford..................... 79

2.4 Commercial Hair Sheep Production in a
Semiarid Region of Venezuela,
Carlos González Stagnaro.................................. 85

2.5 Reproduction and Growth of Hair Sheep
in an Experimental Flock in Venezuela,
Andres Martinez.. 105

2.6　Performance of Barbados Blackbelly Sheep
and Their Crosses at the Ebini Station, Guyana,
G. Nurse, N. Cumberbatch, and P. McKenzie 119

2.7　Hair Sheep Performance in Brazil,
E.A.P. de Figueiredo, E. R. de Oliveira,
C. Bellaver, and A. A. Simplício 125

2.8　Sheep Production in Tobago with Special
Reference to Blenheim Sheep Station,
Raj K. Rastogi, K.A.E. Archibald, and
M. J. Keens-Dumas 141

2.9　Barbados Blackbelly and Crossbred Sheep
Performance in an Experimental Flock in
Barbados, *Harold C. Patterson* 151

2.10　Reproduction and Birth Weight of
Barbados Blackbelly Sheep in the Golden
Grove Flock, Barbados, *G. E. Bradford,*
H. A. Fitzhugh, and A. Dowding 163

2.11　Virgin Islands White Hair Sheep,
Harold Hupp and Duke Deller 171

2.12　A Note on Performance of Barbados Blackbelly
Sheep in Jamaica, *G. E. Bradford,*
A. J. Muschette, Vincent Lyttle, and
David Miller 177

2.13　A History of the Barbados Blackbelly Sheep,
Weslie Combs 179

SECTION 3
WESTERN AFRICA

3.1　Performance of Hair Sheep in Nigeria,
Almut Dettmers 201

3.2　Review of Hair Sheep Studies in
Southwestern Nigeria, *A. A. Ademosun,*
K. Benyi, O. Chiboka, and C. M. Munyabuntu 219

3.3　Djallonké Hair Sheep in Ivory Coast,
Yves M. Berger 227

3.4　A Note on Characteristics of Hair Sheep
in Senegal, *G. E. Bradford* 241

3.5   Sedentary Sheep in the Sahel and Niger Delta
of Central Mali, *R. T. Wilson* 245

SECTION 4
NORTH AMERICA

4.1   Research with Barbados Blackbelly Sheep
in North Carolina, *L. Goode, T. A. Yazwinski,
D. J. Moncol, A. C. Linnerud, G. W. Morgan,
and D. F. Tugman* 257

4.2   The St. Croix Sheep in the United States,
Warren C. Foote 275

4.3   The Barbados Blackbelly ("Barbado")
Breed in Texas, *Maurice Shelton* 289

4.4   Crossbreeding with the "Barbado" Breed
for Market Lamb or Wool Production
in the United States, *Maurice Shelton* 293

4.5   Barbados Blackbelly Sheep in Mississippi,
Leroy H. Boyd 299

4.6   Barbados Blackbelly Sheep in California,
J. M. Levine and G. M. Spurlock 305

Index 313
About the Book and Editors 319

PREFACE

Most publications about sheep are based on experience with wooled sheep in temperate, developed countries. This book, however, is about hair sheep and emphasizes data from tropical, developing countries in western Africa and the Americas.

Our interest in hair sheep was stimulated by the 1977 Winrock International study "The Role of Sheep and Goats in Agricultural Development," which identified hair sheep as an important, but underutilized, animal resource for tropical agriculture. These findings prompted one of us (HAF) to propose a more detailed evaluation of hair sheep and production systems; the other (GEB) had a special interest in sheep breeding, especially the genetics of fertility, and was available to collaborate in the evaluation. We agreed that emphasis should be placed on objective documentation of performance traits and production systems and, most importantly, that attention should be focused on the contributions of developing country specialists with firsthand working knowledge of hair sheep.

Winrock International initiated the project in late 1978, with partial financial support from a USAID Special Activities Grant. The evaluation concentrated on hair sheep populations in the Western Hemisphere and western Africa, the principal source for hair sheep in the Western Hemisphere. We traveled to 15 countries—Mexico, Colombia, Venezuela, Guyana, Brazil, Trinidad and Tobago, Barbados, Dominican Republic, Jamaica, Cameroon, Nigeria, Ivory Coast, Liberia, Mali, and Senegal—as well as to the U.S. and British Virgin Islands. In addition to observing sheep production in these countries, we met with government and university sheep specialists and private flock owners and invited those with performance data on hair sheep to prepare manuscripts for publication in this book.

The book is organized in four sections. The first provides a general description of hair sheep and current production systems, summarizes performance statistics, and makes suggestions for improved breeding, management, and research. The second through fourth sections contain the contributed chapters, which are organized by geographical regions: Middle America–South America, Western Africa, and North America, respectively.

The list of authors of the contributed chapters includes a significant number of the people with experience in the production and management of hair sheep. As these chapters document, their efforts have yielded much useful information about hair sheep. We hope this book will acquaint more people with hair sheep and by so doing attract additional support for future efforts to improve the productivity and efficiency of this important genetic resource.

H. A. Fitzhugh
G. E. Bradford

ACKNOWLEDGMENTS

This book was made possible through the efforts of the contributors, whose names and addresses are listed on the next page. Many of them also served as hosts and guides during our travels. Their contributions to making this project scientifically productive, as well as personally enjoyable, are gratefully acknowledged.

Additional individuals who assisted in evaluation of hair sheep resources included: Barbados—Hugh Jeffers, Lionel Smith, A. Brathwaite, J. and R. Humphreys; Brazil—E. Alves de Moraes; Cameroon—E. D. Tebong, R. D. Ndumbe; Colombia—Juan Salazar, Hugh Scott, Alfonso Naranjo, Edgar Ceballos; Ivory Coast—P. Lhoste, E. Cesenelli, G. Courau; Liberia—L. Jarrett, J. Pan; Mali—A. Cissé, H. N. Le Houerou; Mexico—Leonel Martinez; Nigeria—D. H. Hill, A. K. Mosi, R. Borel (International Livestock Centre for Africa); St. Croix—L. Johansen, H. Moolinaar, M. Christiansen; Senegal—El Hadj Gueye, P. I. Thiongane, I. Diallo; Tortola—David Smith; Trinidad-Tobago—Pascal Osuji, Lawrence Iton, Claude Job, Holman Williams; Venezuela—T. A. Shultz, F. C. Chico.

Part of the financial support for this project was provided by a grant from the U.S. Agency for International Development. Additional assistance in identifying contacts and organizing travel was provided by USAID personnel including Ned S. Raun and J. W. Oxley (Washington), G. Rozelle and T. King (Barbados), E. Wallace (Guyana), J. Cornelius (Liberia), E. Witt (Cameroon), L. Harms (Mali), and W. Thomas (Senegal).

Ian L. Mason suggested valuable contacts and provided a pre-publication copy of Prolific Tropical Sheep, FAO Animal Production and Health Paper 17. He also read early drafts of most chapters in this book and provided many helpful comments. We are particularly grateful to him for his suggestions for clarifying the Sahel-Savanna-Forest nomenclature for West African hair sheep.

Individuals providing photographs used in the book included G. M. Spurlock (Photos 1.1.47 and 4.1.1), J. M. Shelton (Photo 1.1.46), N. Gutierrez (Photos 1.1.7 and 1.1.39), L. Martinez (Photo 1.1.2), and Ned S.

Raun (Photo 1.1.34). Photo 1.1.43 was provided by Heifer Project International. The other photographs were provided by us.

The survey was accomplished while one of us (GEB) was on sabbatic leave from the University of California, Davis. Appreciation is expressed to the university for this support of the project.

We also express our special appreciation to the following members of the Winrock International staff: translations—N. Gutierrez, E. Ospina, and A. Martinez; secretarial assistance—Irene Osment, Sarah Cragar, Libby Fowler, Tammy Chism, Darlene Galloway, and, especially, Shirley Zimmerman; proofreading—Essie Raun. Throughout the project, Beth Henderson provided careful attention to the multitude of technical and editorial details inherent in a book of this kind. Jim Bemis brought his considerable experience and good humor to bear on the production editing problems of the book.

Lynn Arts, Don Meyer, and others at Westview Press made important editorial contributions to the clarity of the book, and their efforts are sincerely appreciated.

To these people and all the others who helped us in ways small and large, we express our gratitude.

H.A.F.
G.E.B.

THE CONTRIBUTORS

Ademosun, Akin A. Department of Animal Science, University of Ife, Ile-Ife, Nigeria.

Archibald, K.A.E. Department of Livestock Science, University of West Indies, St. Augustine, Trinidad.

Bellaver, Claudio. Centro Nacional de Pesquisa de Caprinos, EMBRAPA, Sobral, Ceará, Brasil.

Benyi, K. Department of Animal Science, University of Ife, Ile-Ife, Nigeria.

Berger, Yves M. Centre de Recherches Zootechniques, Bouaké, Ivory Coast.

Bodisco, V. Instituto de Investigaciones Zootecnicas, CENIAP, Maracay, Venezuela.

Boyd, Leroy H. Department of Animal Science, Mississippi State University, Mississippi State, MS.

Bradford, G. Eric. Department of Animal Science, University of California, Davis, CA.

Camacho D., Rafael. Caja Agraria, Bogotá, Colombia.

Chiboka, O. Department of Animal Science, University of Ife, Ile-Ife, Nigeria.

Combs, Weslie. Consultant in Livestock Development, Box 2300 Station D, Ottawa, Ontario, Canada, K1P 5W4.

Cumberbatch, N. Ministry of Agriculture, Georgetown, Guyana.

De Alba, Jorge. Centro de Adiestramiento y Mejoramiento, Asociación Mexicana de Producción Animal, A. C., Tampico, Mexico.

Deller, Duke. Veterinary Services, Department of Agriculture, St. Croix, U.S. Virgin Islands.

Dettmers, Almut. Department of Animal Science, Faculty of Agriculture, University of Ibadan, Ibadan, Nigeria.

Diallo, I. Centre Recherche Zootechnique, Senegal.

Dowding, A. Golden Grove Estates, Barbados.

de Figueiredo, Elsio. Centro Nacional de Pesquisa de Caprinos, EMBRAPA, Sobral, Ceará, Brasil.

Fitzhugh, H. A. (Hank). Winrock International, Morrilton, AR.

Foote, Warren C. Department of Animal Science, Utah State University, Logan, UT.

Fuenmayor, C. Instituto de Investigaciones Zootecnicas, CENIAP, Maracay, Venezuela.

Gonzalez Padilla, Everardo. Instituto Nacional de Investigaciones Pecuarias, SARH, Mexico DP, Mexico.

Gonzalez-Reyna, A. Centro de Adiestramiento y Mejoramiento, Asociación Mexicana de Producción Animal, A. C., Tampico, Mexico.

González Stagnaro, Carlos. Facultad de Agronomía, Universidad de Zulia, Maracaibo, Venezuela.

Goode, Lemuel. Department of Animal Science, North Carolina State University, Raleigh, NC.

Gueye, El Hadj. Centre de Recherches Zootechniques de Kolda, Instituto Senegalais de Recherches Agricoles, Kolda, Senegal.

Hupp, Harold. Virgin Islands Agricultural Experiment Station, St. Croix, U.S. Virgin Islands.

Keens-Dumas, M. J. Ministry of Agriculture, Lands and Fisheries, Trinidad and Tobago.

Levine, J. M. Department of Animal Science, University of California, Davis, CA.

Linnerud, A. C. Department of Experimental Statistics, North Carolina State University, Raleigh, NC.

Lyttle, Vincent. Agriculture Development Corporation, Kingston 5, Jamaica.

Martinez, Andres. Winrock International, Morrilton, AR.

Mazzarri, G. Instituto de Investigaciones Zootecnicas, CENIAP, Maracay, Venezuela.

McKenzie, Patrick. Ministry of Agriculture, Georgetown, Guyana.

Miller, David. Agricultural Development Corporation, Kingston 5, Jamaica.

Moncol, D. J. Department of Veterinary Science, North Carolina State University, Raleigh, NC.

Morgan, G. W. Department of Poultry Science, North Carolina State University, Raleigh, NC.

Munyabuntu, C. M. Department of Animal Science, University of Ife, Ile-Ife, Nigeria.

Muschette, A. J. Agricultural Development Corporation, Kingston 5, Jamaica.

Nurse, G. Ministry of Agriculture, Georgetown, Guyana.

de Oliveira, Ederlon R. Centro Nacional de Pesquisa de Caprinos, EMBRAPA, Sobral, Ceará, Brasil.

Pastrana B., Rodrigo. Instituto Colombiano Agropecuario, Bogotá, Colombia.

Patterson, Harold C. Ministry of Agriculture, Barbados.

Rastogi, Raj K. Department of Livestock Science, University of West Indies, St. Augustine, Trinidad.

Reverón, Angel E. Instituto de Investigaciones Zootecnicas, CENIAP, Maracay, Venezuela.

Shelton, J. Maurice. Agricultural Research and Experiment Center, Texas A&M University, San Angelo, TX.

Simplício, Aurino Alves. Centro Nacional de Pesquisa de Caprinos, EMBRAPA, Sobral, Ceará, Brasil.

Spurlock, G. M. Department of Animal Science, University of California, Davis, CA.

Thiongane, P. I. Centre Recherche Zootechnique, Senegal.

Tugman, D. F. Upper Mountain Research Station, Laurel Springs, NC.

Valencia Zarazúa, Mario. Centro Experimental Pecuario Mococha. INIP SARH, Mérida, Yucatan, Mexico.

Wilson, R. T. International Livestock Centre for Africa, Programme du Sahel, Bamako, Mali.

Yazwinski, T. A. Department of Animal Science, University of Arkansas, Fayetteville, AR.

SECTION ONE:
GENERAL OVERVIEW

1.1.1 Barbados Blackbelly ewe with newborn twins, Barbados.

HAIR SHEEP: A GENERAL DESCRIPTION

G. E. Bradford, *University of California, Davis, U.S.A.*
H. A. Fitzhugh, *Winrock International, U.S.A.*

Sheep share with other ruminants the ability to convert fibrous, often low-quality feedstuffs into valuable, life-sustaining products for human use, such as meat, milk, wool, and skins (1). This advantage has been recognized since sheep were domesticated in southwestern Asia over 10,000 years ago. Domestic sheep, *Ovis aries*, are believed to be descendents of wild species that still exist and are interfertile with domestic sheep. These wild species include the Asian and European Mouflon (*O. orientalis* and *O. musimon*), the Argali (*O. ammon*), and the Urial (*O. vignei*). A fifth species, the Bighorn of North America (*O. canadensis*), was never domesticated, but has the same chromosome number (2N = 54) and appears to be interfertile with *O. aries* (2).

Hair sheep, the subject of this book, have evolved under the influence of selection by both nature and man. Unlike their better-known wooly relatives, they have hair coats similar to those of cattle and goats. (Other names for hair sheep are "Pelibuey" in Cuba and Mexico, and "Pelo do Boi" in Brazil, meaning hair of the ox.) Because of this similarity in coat types, hair sheep may be mistaken for goats. However, goats belong to the genus *Capra*, have a diploid chromosome number of 60, and are not interfertile with sheep. A prosaic, but obvious, distinction is that a sheep's tail hangs down, whereas a goat's tail is usually held erect.

Although neither as numerous nor as economically important as wool sheep, hair sheep are well adapted to tropical environments, an important advantage in many developing countries where hair sheep are the principal, or only, sheep. For centuries, agriculturists in tropical regions have found hair sheep to be both a dependable source of food and a ready producer of extra income. The principal products from hair sheep flocks are animals for sale and meat for family consumption. Hides are an important by-product, but fiber is not. Ewes may be milked, but this practice is infrequent. Sheep manure is used to fertilize crops.

Hair sheep are usually found at the lower altitudes in the tropics. They account for most of the 90 million sheep in tropical Africa, the 2 million sheep in the Caribbean region, and the approximately 6 million sheep in northeast Brazil. Hair sheep also are found in southern India and southeastern Asia (3). In total, we estimate that hair sheep comprise 7 to 10%—about 100 million head—of the world's 1.1 billion sheep population.

Even though hair sheep currently produce only a small fraction of the world meat supply, demand for sheep meat will likely increase to meet the escalating needs of human populations in developing countries. There is additional potential for hair sheep in regions such as the southern United States, where wooled sheep are poorly adapted and where the cost and effort of shearing are not justified by the value of the harvested wool.

Hair sheep merit serious efforts to develop their potential to meet human needs. An important first step is the thorough characterization of this important genetic resource.

COAT CHARACTERISTICS OF HAIR SHEEP AND WOOL SHEEP

Three types of fibers—wool, kemp, and hair—are found in the coats of sheep (Figure 1.1.1). These fibers differ in diameter, and some have a medulla (hollow or mostly hollow core). Medullae may extend the length of the fiber or be discontinuous. They are further classified as latticed or nonlatticed, based on the appearance of the internal structure of the medulla (4, 5).

Wool fibers are the finest, varying from approximately 15 to 40 μm in diameter. Fine wool is unmedullated, but coarser wool may have a narrow, usually discontinuous, nonlatticed medulla.

Kemp fibers are the coarsest of the three types, approximately 100 μm in diameter. Their latticed medulla usually extends the length of the fiber and constitutes over 65% of the cross-sectional area of the fiber.

Hair fibers, also called heterotypes, are intermediate between wool and kemp in diameter and relative cross-sectional area of the medulla. The usually nonlatticed medulla may be interrupted along the length of the fiber.

Fibers are produced from primary and secondary skin follicles. Primary follicles are larger and include a sweat gland and erector muscle. Secondary follicles are more numerous, occur in groups around the primaries, and lack the gland and muscle structures. Coarse, medullated fibers are commonly produced from primary follicles, while the finer, often unmedullated wool and heterotype fibers are produced from secondary follicles.

Wild sheep have an outer coat of coarse hair over an undercoat of fine

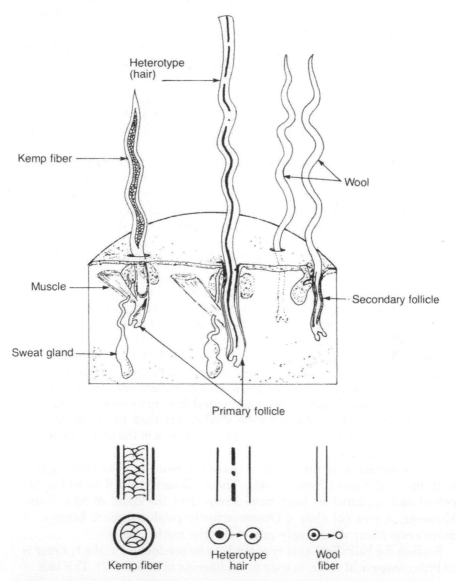

FIGURE 1.1.1 Illustrations of fiber types: kemp fiber with continuous wide diameter, latticed medulla; hair or heterotype fiber with fragmented, narrow, nonlatticed medulla; and wool with solid core.

wool. The ratio of secondary to primary (S/P) follicles is low, generally 2/1 or 3/1. Both hair and wool are shed in the spring (Northern Hemisphere); some shedding of hair and kemp occurs in the autumn as well.

Coat types of domesticated wool sheep have been greatly changed through selection. Coarse-wool sheep still produce medullated fibers from primary follicles and have a relatively low S/P ratio (3/1 to 4/1), with considerable variation in fiber diameter within the fleece. Nevertheless, coarse-wool sheep produce fewer kemp and more wool fibers than do their wild ancestors. Fine-wool sheep produce wool fibers from the primary follicles that are similar in diameter and character to those produced by the secondary follicles; the S/P ratio is very high (up to 20/1) and follicle density is also high.

The hair sheep coat has been likened to that of their wild ancestors; namely coarse outer hairs over fine wool. Usually, however, the undercoat of wool is sparse (or even nonexistent) and not evident on visual inspection. Most hair is typically 1 to 3 cm in length, although some is considerably longer. For example, adult rams may have a mane and throat ruff (Photo 1.1.13).*

There are reports that growth of the wool undercoat is stimulated when hair sheep are moved from tropical to cool temperate climates (6); however, the degree and extent of wool growth remains undocumented.

Burns (7) compared the coat and skin characteristics of Nigerian hair sheep, imported Merinos, and their F_1 crosses. The hair sheep had a lower density of primary follicles and a lower S/P ratio (less than 4/1) than the Merinos. The coat type of F_1 animals resembled that of the hair sheep parent more than that of the wool sheep parent. All F_1 fleeces contained many short kemp fibers, and the wool fibers were extremely fine and soft. Fleece shedding generally began on the mid-ventral line and proceeded dorsally; fibers on the withers and rump were shed earlier than those on the mid-back. The last part to be shed was a saddle-like area in the mid-back region (e.g., Photo 1.1.41).

Our observations of the fleeces of crosses of wool and hair sheep agreed with those of Burns (Photo 1.1.41). These "fleeces" tended to be heavily cotted and appeared to have more kemp than the coats of hair sheep. However, it was not clear if crosses actually produced more kemp or if more kemp fibers were simply retained in the matted fleece.

Because the hair sheep coat type tends to be predominant, the F_1 fleece is of little commercial value, in spite of the fineness of its wool (7). This lack of

---

*Photos do not necessarily appear in the text near their references because many of them are in color and have been grouped together for printing on special paper. The color section appears immediately after p. 48.

value, along with the probable maladaptive aspects of a matted coat, raises serious questions about the advisability of crossing hair sheep with wool sheep, particularly in the tropics.

Ryder and Stephenson (4) considered the similarities between the coats of hair sheep and of wild sheep as evidence that hair sheep are more primitive than wool sheep. Zeuner (8), on the other hand, considered the coat of hair sheep to be a specialized adaptation, resulting from selection on the primitive genotype. Regardless of the evolutionary history, the absence of a thick wool coat is adaptively advantageous for hair sheep in the tropics and, when compared with the qualities of "improved" breeds, should be regarded as an asset to be exploited, rather than as an inferior characteristic.

TYPES AND BREEDS

The types and breeds described in this chapter (and throughout the book) are those found in western Africa and the Americas (Figure 1.1.2). These sheep are the thin-tailed type, except for the Blackhead Persian. Epstein (9) cited evidence of long-legged, thin-tailed hair sheep in northern Africa (Egypt) in the third millennium b.c., presumably of southwest Asian origin. Although there is no documentation, it seems probable that thin-tailed hair sheep could have migrated from northeastern Africa through north-central Africa, eventually reaching western Africa.

Hair sheep were introduced to Brazil and the West Indies in the seventeenth century A.D., probably from western Africa.

West Africa

Hair sheep in West Africa have been classified into two types: a larger, longer-legged, lop-eared type found in the drier Sahelian zones (Photo 1.1.16) and a smaller, horizontal-eared type found in the humid Forest and Savanna zones (Photo 1.1.19). Epstein (9) referred to the larger, long-legged type as the Savanna and the smaller type as the West African Dwarf. These names are confusing on two counts: the larger, long-legged sheep are primarily associated with the Sahelian zone and the smaller types are not true dwarfs.

A study published by the International Livestock Centre for Africa (10) divided woolless thin-tailed sheep in West Africa into two types: Sahel and Forest or Savanna. These type designations refer to the ecozones in which each type is commonly found (Figure 1.1.3). For consistency, we have used three general names: Sahel, Savanna, and Forest; however, Forest and Savanna types are generally discussed together. In francophone countries, the Forest and Savanna types are generally called Djallonké and Fouta Djallon; in anglophone countries, they are often called West African

value, along with the probable maledetate aspect, is a distributor, based
accrue features about their availability of expense and sheep with a of
sheep, particularly in the future.

Ryder and Stephens' (references...) the significance between the ...

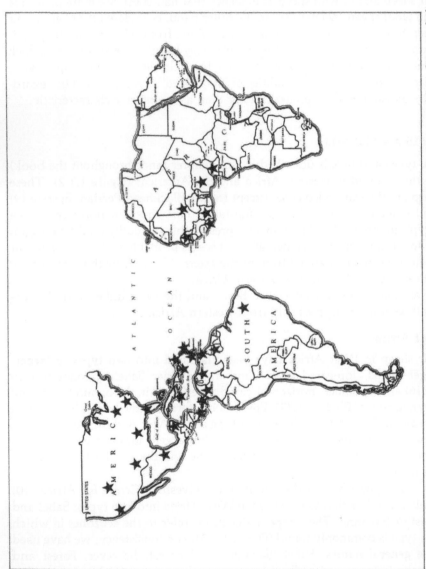

FIGURE 1.1.2 Stars indicate countries visited by Bradford and Fitzhugh and sites providing data on hair sheep.

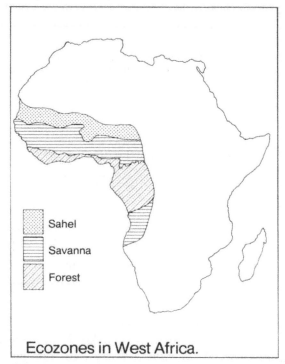

Ecozones in West Africa.

FIGURE 1.1.3 Sahel, Savanna, and Forest types of sheep are found predominantly in ecozones of the same names.

Dwarf. Local names by which they are known vary widely (Table 1.1.1). Terminology in contributed chapters was left to the discretion of the respective authors.

Principal phenotypic differences between the Sahel and Forest-Savanna types are described in Table 1.1.2. Photographs of the respective types also indicate some of these differences (e.g., Sahel type: Photos 1.1.2, 1.1.11, 1.1.16; Forest-Savanna type: Photos 1.1.19, 1.1.21).

Size traits vary greatly, especially body weight, which is subject to environmental influences on degree of fatness. Sahel types are taller than the Savanna type and both taller and heavier than the Forest type. Sahel sheep are generally associated with nomadic or transhumant production systems. Their long legs facilitate travel over long distances and browsing on trees and shrubs, a principal feed source through much of the year. Mature weight of the Sahel-type ewes we saw in Senegal was 36 kg (12), which is within the range of mature weight for Savanna sheep reported by Mason (13). However, these Sahel sheep were in thin condition, and we estimated

TABLE 1.1.1. SOME GENERAL AND LOCAL NAMES FOR HAIR SHEEP
IN WEST AFRICA

SAHEL TYPE

Mason (3):   Fulani with French equivalents Peul, Peulh, Peul-
Peul, Foulbé; West African Long-legged and Guinea Long-
legged with local variants Bornu, Samburu; Toronké, Ouda
(Uda), Yankasa.

Doutressoulle (11):   Maure, Peul, Tuareg, and Bali-Bali (in
Nigeria).

Ibrahim (11):   Balami and Targui in Niger.

Dia (11):   Peul-Peul and Touabire in Senegal.

FOREST AND SAVANNA TYPES

Mason (3):   Djallonké, Fouta Djallon, Kirdi, Kirdimi or Lakka,
Nigerian Dwarf, West African Dwarf, West African Maned.

ILCA (10):   Djallonké, Fouta Djallon, Southern, Guinean.

---

that in good condition their mean weight would have exceeded 40 kg. Sahel-
type ewes observed at markets (e.g., in Bamako, Mali) were estimated to
weigh up to 50 to 60 kg. We saw one mature Sahel ram judged to weigh
over 100 kg.

Forest-Savanna sheep are more compact than the Sahel sheep. Epstein (9)
called the Forest type "dwarf" on the basis of its small mature height, which
is typically less than 60 cm at the withers. We prefer not to use the term
"dwarf" because the Forest sheep that we observed were small, but not
achondroplastic. However, some goats in the region are achondroplastic
(Photo 1.1.3) and are well described by the name, West African Dwarf.

Sheep in the savanna zone are generally taller and heavier than sheep in
the coastal forest zone, perhaps because of less stress from climate and
disease. Also, sheep raised at higher altitudes tend to be larger than those
raised at lower altitudes (10).

Characteristics that distinguish the Sahel and Forest-Savanna types include trypanotolerance, mane and throat ruff, tail length, and ear length-
shape-carriage (Table 1.1.2). Wattles (also called throat tags, lappets, toggles, tassels, and *Appendices colli*) are common among Sahel sheep but less
so among the Forest-Savanna type. Wattles are inherited as a characteristic
determined by a single dominant, presumably autosomal, gene (9).

Sahel sheep are generally white or white and brown (Photos 1.1.2,
1.1.16). Forest-Savanna sheep in the region from Senegal to Nigeria are

TABLE 1.1.2. CHARACTERISTICS OF SAHEL, FOREST AND SAVANNA
TYPES OF HAIR SHEEP IN WEST AFRICA[a]

	Sahel Type	Forest and Savanna Types
Weight of mature females	Over 35 kg	Forest,[b] 20 to 30 kg; Savanna, 30 to 40 kg
Withers height of mature females	Over 60 cm[c]	Forest,[c] 40 to 55 cm; Savanna, 55 to 65 cm
Ears	Long, pendulous	Short, horizontal
Wattles	Common	Rare
Horns Males	Horned; frequently long and spiralled (screw-shaped) horns	Usually short, crescent shaped horns, some polled
Females	Usually polled	Polled, some scurs
Tail length	Usually extends below hocks	Usually above hocks
Mane and/or throat ruff on males	Absent	Usually present
Color	Usually white or white and brown	Senegal to Nigeria – white, black, white and black, some browns. Cameroon to Angola – white, brown (various shades), white and brown, blackbelly, some black
Trypanotolerant[c]	No	Yes

a Authors' observations unless otherwise noted.
b Mason (13)
c ILCA (10)

commonly white with black spotting, although there are some solid white
and black sheep (Photos 1.1.20, 1.1.21, 1.1.23).

Red, tan, brown (solid or spotted), and blackbelly patterns are found in
the Senegal-Nigeria region (Photo 1.1.22); however, these colors are encountered more frequently in Cameroon, Gabon, Angola, and other
western-central African countries (author's observations; 13). These are the
colors and patterns most common in the Americas, which has bearing on
speculations about the source of hair sheep in the Americas.

The Sahel-type rams that we saw were maneless and horned; some ewes
were horned; long, thin tales often extended below the hocks (Photo 1.1.18).
The long spiral horn shape was most common (Photo 1.1.2).

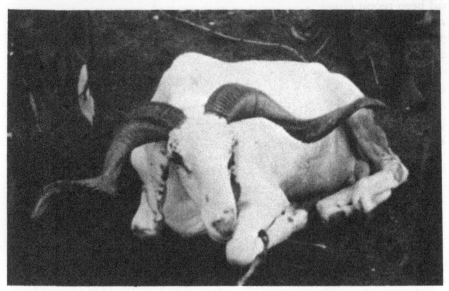

1.1.2 Aged Sahelian ram, Ivory Coast.

1.1.3 West African Dwarf doe and Forest-type ewe, in village near Ibadan, Nigeria.

Most of the Forest-type rams we saw were horned, but some polled rams were also observed (Photo 1.1.13); we saw no horned ewes. The typical horn shape is a single curve or crescent (Photo 1.1.19). Typically, Forest-Savanna rams have a mane and throat ruff of long, coarse hair (Photos 1.1.13, 1.1.20). Tails rarely extend below the hocks (Photos 1.1.19, 1.1.25).

General Comments. Wool sheep have been introduced into most West African countries. However, except for experimental flocks in which there had been recent crossbreeding, the sheep that we saw in the humid zones were free of wool. Perhaps the imported wool sheep never reached the village flocks or, if they did, the wool genotype was eliminated by natural selection for adaptation to local climate and disease conditions. Certainly, a noteworthy characteristic of the Forest sheep is that they appeared healthy and adapted to the humid climate and to local health problems such as trypanosomiasis.

The general absence in the humid coastal zone of sheep showing Sahel-type characteristics is somewhat surprising since Sahelian sheep are routinely marketed in the more densely populated, higher-rainfall areas nearer the coast. Although many sheep are now transported by truck, the traditional practice has been to walk them to the urban markets. Perhaps little intermating of Sahelian sheep with local sheep has occurred because the trails avoid the local villages. This possibility is supported by the observation (14) that sheep near the major urban markets in Ivory Coast exhibit some Sahel-type characteristics, whereas sheep in the smaller village flocks do not. Another possibility is that offspring of the crosses lack resistance to trypanosomiasis and other local diseases and thus are eliminated by natural selection. Certainly, limited data indicated mortality rates of Sahelian-Djallonké cross lambs were appreciably higher than those of Djallonké lambs at the Centre de Recherches Zootechniques, Bouaké, Ivory Coast.

The Americas

Hair sheep populations are found on many Caribbean islands, in Central and South American countries around the Caribbean basin, and in northeastern Brazil. Both Barbados and the Virgin Islands have exported hair sheep to the United States.

American hair sheep more closely resemble the Forest-Savanna type than the Sahel type of western Africa, although they are generally larger than the Forest type. The fat-rumped Blackhead Persian resembles neither and is clearly of different origin.

Historical evidence (13, 15) and observed phenotypic similarities support the notion of a West African origin for ancestors of American hair sheep. These ancestors probably arrived by ship in the sixteenth and seventeenth centuries during the height of the slave trade.

Comparisons of photos 1.1.10 through 1.1.15 indicate many similarities between present-day West African and American hair sheep, even though these photos were not taken with prior intent to demonstrate similarities.

Mason (13) considered the higher frequency of tan, brown, or red sheep in the American population as evidence that principal importations had come from the central portion of West Africa (Cameroon to Angola) rather than from the coastal region from Nigeria to Senegal, where black sheep are most frequent. He further pointed out that American hair sheep are more similar in size to the Savanna type than to the smaller Forest type found along the west African coast. However, the larger size of American hair sheep may be a consequence of development in a more favorable environment, including better nutrition and fewer health problems.

Hair sheep in both tropical Africa and America are fertile throughout the year, except for periods of nutritional and/or lactational anestrus.

An obvious difference between the present-day populations is in the frequency of horned animals: most West African rams and some females are horned, whereas practically all American hair sheep are polled. Mason (13) suggested that polled animals were preferred for transoceanic voyages in the confined space available on shipboard and that subsequent selection in the Americas favored the polled type.

Characteristics common to most American hair sheep include:

- Mature weights of 30 to 40 kg for ewes and 45 to 60 kg for rams
- Compact conformation
- Thin tails extending to hocks or shorter
- Small ears carried horizontally
- Rams and ewes usually polled
- Pronounced mane and throat ruff on many rams although less common for Pelibuey and rare for Brazilian breeds.

White and shades of brown from tan to red to dark brown are the most frequent coat colors. Common patterns include solids (Photos 1.1.12, 1.1.31, 1.1.34), various types of spotting (Photos 1.1.33, 1.1.35, 1.1.38), and the distinctive blackbelly pattern (Photo 1.1.14).

We saw the greatest variety of colors and patterns in the flock of Pelibuey at Mococha, Mexico (Photo 1.1.35). Solids included white, a full range of browns from light beige to "red" to deep chocolate, and (rarely) black. Spots ranged in size from those that covered half the animal to small specks, called "mosqueado." Variations included white on brown; brown on white; blackbelly; golondrina (reverse blackbelly pattern—light tan undercolor and facial stripes on dark body); and payaso or harlequin, the blackbelly pattern on a spotted or roan body (Photo 1.1.28).

Analysis of data from the Mococha flock (Table 1.1.3) suggests that white and brown colors are determined by different alleles of a single autosomal gene, with white being dominant to brown. Explanations for the unexpected white progeny from brown × brown matings include mistaken parentage and misclassification of body color due to spotting (e.g., parents may have had brown spots on a white body, rather than the recorded white spots on a brown body).

Subpopulations of American Hair Sheep. A strong argument can be made for considering all American hair sheep (except the fat-rumped Blackhead

TABLE 1.1.3. SUMMARY OF PROGENY COLORS BY TYPE OF MATING
IN AN EXPERIMENTAL FLOCK OF PELIBUEY SHEEP IN
MEXICO.

Type of mating	Progeny color	Observed number
White x White	White	206
	Brown	60
White x Brown	White	86
	Brown	89
Brown x Brown	White	23
	Brown	229

Source: M. Valencia Z. (personal communication)

Persian) as a single genetic type. Given the general similarity among their probable ancestors (Forest-Savanna types) throughout western Africa, the similarity among populations of American hair sheep is not surprising.

Over time, however, the combined effects of geographical isolation, of local selection for special traits, and of outcrossing to European breeds may have brought about real genetic differences among subpopulations of American hair sheep. For example, the Barbados Blackbelly tends to be more ectomorphic and, most importantly, more prolific than the norm for American hair sheep.

The considerable within-group variation for distinguishing phenotypes, such as coat color and pattern, makes any classification tentative at best. In the procedure we have followed, breeds and types are listed by their most commonly used names. These American breeds and types are further combined into six groups: Barbados Blackbelly, Virgin Islands, Pelibuey–Africana–West African, Blackhead Persian, and the two Brazilian breeds Morada Nova and Santa Inês. These groupings are largely based on our subjective evaluation of phenotypic similarities, but also take note of probability of interbreeding based on geographical proximity and recorded importations.

Barbados Blackbelly. Probably the best known of the Caribbean breeds (6, 16), the Barbados Blackbelly differs from other American hair sheep in several respects: the distinctive color pattern, ectomorphic conformation, and higher prolificacy. The blackbelly pattern was described by Rastogi et al. (16):

Body colour varies from light to dark reddish-brown (tan) with very conspicuous black underparts. The black colouration covers the lower jaw, the chin, throat, breast, entire belly, auxillary and inguinal regions, and inner sides of the legs, and extends as a narrow line along the underside of the tail nearly to the tip. On the outer side of each leg, the paler colour persists dorsally only as a restricted and more or less broken stripe. The inner surface of the ear is black, and there is a conspicuous black stripe on the face above and anterior to each eye and to the tip of the muzzle. In the adult male, the occipital area immediately behind the horn bases is also black. Where the hair is short, as on the breast and belly, the black area is sharply delimited, but in the longer hair of the outer sides of the thighs and on the mane of the male the transition from black to pale colour is more gradual. The colour of the back and sides is reddish-brown, which becomes paler on the face, the sides of the neck, and the flanks. A white spot is found below and slightly in front of each eye and sometimes another smaller white spot above it. The tip of the tail may occasionally be white.

The blackbelly pattern is termed badger-face by geneticists whether the back is white or tan. It is in the agouti series and appears to be recessive to self-colour white or tan but dominant to black (17).

Sheep from Barbados have been exported to many Caribbean, Central and South American countries where they are often maintained in closed flocks (Photos 1.1.44, 1.1.45). A few Barbados Blackbelly sheep have been imported to the United States, where they have been crossed to Rambouillet, Dorset, other wooled breeds, and the wild Mouflon (18). As a consequence, many "Blackbelly" types in the United States are horned, produce some wool, and do not show the distinctive color pattern (Photos 1.1.46, 1.1.47).

Virgin Islands. White is the predominant and apparently preferred color for sheep in the U.S. and British Virgin Islands, leading to the common name of Virgin Islands White (Photo 1.1.12). However, tan, other shades of brown, and some black markings are seen among sheep on these islands (Photo 1.1.33). Since white is apparently dominant to brown, continuous selection for white could easily account for the present-day incidence of white sheep.

Similar white sheep are found through the Lesser Antilles, the Bahamas (19), and other Caribbean islands. These white sheep often are considered by local producers to be a different type than local brown sheep. However, we found no objective basis for this distinction.

Preliminary evidence summarized in Chapter 1.2 suggests that Virgin Islands sheep are more prolific than other American hair sheep, except the Barbados Blackbelly.

Sheep imported to the state of Maine from St. Croix, one of the U.S. Virgin Islands, were combined with Suffolk and Wiltshire Horn (Photo 1.1.42) breeds to develop the Katahdin breed (Photo 1.1.43). More recently, sheep imported to the United States from St. Croix have been used in experiments reported in Chapter 4.2.

1.1.4 Pelibuey ram, Cuba.

Pelibuey–Africana–West African. This grouping includes sheep of similar size, color, and performance from a number of countries in the Caribbean region.

Pelibuey sheep from Cuba (Photo 1.1.4) are known to have been exported to Mexico and to the Dominican Republic (13). Exports to other countries near Cuba, such as Haiti and Jamaica, are probable. Mason (13) reported that tan, white, and tan and white are the most frequent coat colors seen in Cuba, with "red" being the preferred color. Other common names include Cubano Rojo and, in Mexico, Tabasco and Peliguey (Photos 1.1.28, 1.1.35).

The Africana sheep in Colombia and Venezuela are usually brown, ranging in shade from tan to brown and cherry-red to dark red (Photos 1.1.5, 1.1.29, 1.1.34). Other names are Pelona, Camura, and Red African (Rojo Africana) (20).

Tan or brown sheep (except those with the blackbelly pattern) are usually

1.1.5 Reddish brown Africana rams, Venezuela.

called "West African" in English-speaking Caribbean countries (Photo 1.1.6).

Brazilian Breeds. The common type of hair sheep found in Northeast Brazil is called the Pelo do Boi or, sometimes, Ovino Deslanado. These sheep are similar in appearance to the Pelibuey–West African types of the Caribbean region. The more common colors are red, brown, and white with various spotting patterns. Typically, rams have neither manes nor throat ruffs. The Morada Nova (red or white) breed (Photo 1.1.31) has been established from the Pelo do Boi. It is similar in appearance (ear and tail type, size, and conformation) to most Caribbean hair sheep.

Another Brazilian breed, the white or red Santa Inês, differs from the Morada Nova in many respects. Santa Inês are larger, longer-legged, Roman-nosed, and lop-eared (Photo 1.1.7). These characteristics of the Santa Inês generally are thought to have been derived from crossing the Morada Nova with the coarse-wooled Italian breed, Bergamasca (21); however, there are striking similarities between the Santa Inês and Sahelian types of hair sheep (compare Photos 1.1.10 and 1.1.11).

Blackhead Persian. Small numbers of these fat-rumped sheep (Photo 1.1.37) are found in several Caribbean-basin countries, including Trinidad and Tobago, Venezuela, Colombia, and in Brazil where they are called Somali Brasileira (Photo 1.1.32). Their color pattern (white body and black head and neck, with the two colors sharply demarcated), fat rump, short legs, compact conformation, and low twinning rate clearly distinguish them from other American hair sheep.

The Blackhead Persian was introduced to the Caribbean from South

1.1.6 "West African" ewe tethered in common grazing land, Trinidad and Tobago.

Africa where the breed had developed from a sample of three animals brought from eastern Africa in 1868 (9). The Lowlands Estate flock in Tobago has been the source of breeding stock for many, perhaps all, of the Blackhead Persian flocks now found in the Caribbean region. It is commonly believed that the "pure" Blackhead Persian flocks of the Western Hemisphere are highly inbred; however, the presence of varying amounts of wool on some animals in most flocks suggests that there has been outcrossing to wool sheep since their introduction.

COMPARISON OF HAIR SHEEP AND WOOLED SHEEP

Generally accepted norms for sheep performance have been based almost entirely on wooled breeds developed and evaluated in temperate environments. Our observations and information in this book provide a basis for some general comparisons of hair sheep against these norms.

Compared to wooled sheep in temperate environments, hair sheep in tropical environments:

- are smaller, and grow more slowly
- are early maturing and may lamb at younger ages

- are similar for most fertility traits with similar interbreed variation, except that with adequate nutrition hair sheep are fertile throughout the year (wooled sheep show marked seasonal differences in fertility)
- have higher mortality rates
- have lower dressing percentages and lighter carcass weights but are usually older at slaughter

These comparisons have limited utility because genetic differences between hair and wooled sheep are confounded with differences between tropical and temperate environments and management practices.

When hair sheep and wooled sheep are compared in tropical environments, hair sheep:

- are much better adapted to tropical conditions
- are more fertile and have higher survivability at all ages
- have pre- and post-weaning growth rates similar to wool sheep

A limited amount of information is available on the performance of hair sheep, and on hair sheep crosses with wooled breeds in the United States. This information suggests that hair sheep:

- are smaller and have lower growth rates
- have a considerably longer breeding season and tend to be fertile throughout the year even at temperate latitudes
- lambs (especially offspring of crosses) have better survivability and show superior resistance to internal parasites
- are more prolific than most breeds in the United States, except the Finnish Landrace
- have similar dressing percentages but a higher percentage of kidney fat when slaughtered at 40 to 45 kg

An Important Genetic Resource

Populations of hair sheep tend to be considerably more heterogeneous than are temperate-zone breeds. The following chapters document the existence of sheep that differ markedly from temperate breeds and provide evidence of the extent to which subgroups differ from each other. The heterogeneity within groups, whether attributable to relatively recent crossing or simply to lack of prior directional selection, offers an unusual opportunity for effective selection to improve productivity. This heterogeneity, coupled with the adaptation of the sheep to tropical environments through generations of natural selection, makes the hair sheep a genetic resource of great potential for tropical agriculture. Opportunities for realizing this potential are discussed in the next chapter.

REFERENCES

1. The major domestic ruminants are cattle, sheep, goats, water buffalo, and the camelidae. Ruminants have a multicompartmented stomach that facilitates digestion of fibrous feedstuffs such as forages and crop residues. The largest compartment, the rumen, functions as a large fermentation chamber in which millions of symbiotic microorganisms break down complex cellulosic and other materials into relatively simple nutrients that can be utilized to support animal productivity. Fitzhugh, H. A., H. J. Hodgson, O. J. Scoville, T. D. Nguyen, and T. C. Byerly. 1978. *The role of ruminants in support of man.* 136 pp. Morrilton, Arkansas: Winrock International.
2. Brooke, C. 1978. *The domestication of sheep.* Pp. 18–20. Logan, Utah: Int. Sheep and Goat Inst., Utah State University.
3. Mason, I. L. 1969. *A world dictionary of livestock breeds, types and varieties.* 268 pp. Tech. Comm. No. 8 (revised) of the Commonwealth Bur. of Animal Breeding and Genetics, Edinburgh. Commonwealth Agr. Bureaux, Farnham Royal, Bucks., England.
4. Ryder, M. L. and S. K. Stephenson. 1968. *Wool growth.* London and New York: Academic Press.
5. Von Bergen, W. and H. R. Mauersberger. 1948. *American wool handbook.* New York: Textile Book Publishers, Inc.
6. Patterson, H. C. 1976. *The Barbados Blackbelly sheep.* Bulletin of Barbados Ministry of Agriculture, Science and Technology, Bridgetown, Barbados.
7. Burns, M. 1967. The Katsina wool project. I. The coat and skin histology of some northern Nigerian hair sheep and their Merino crosses. *Tropical Agriculture* (Trinidad) 44:173–192.
8. Zeuner, F. E. 1963. *A history of domesticated animals.* London: Hutchinson.
9. Epstein, H. 1971. *The origin of the domestic animals of Africa. Vol. II.* New York, London, and Munich: Africana Publishing Company.
10. ILCA. 1979. *Trypanotolerant livestock in West and Central Africa. Volume 1: General Study.* ILCA Monograph 2. Addis Ababa, Ethiopia: International Livestock Centre for Africa.
11. Dia, Papa Ibrahim. 1979. L'élevage ovin au Sénégal; situation actuelle et perspectives d'avenir. Thèse, Ecole Inter-Etats des Sciences de Médicines Vétérinaires, Dakar, Sénégal.

    Doutressoulle, C. 1947. *L'élevage en Afrique Occidentale Française.* Paris: Maisonneuve et Larose.

    Ibrahim, Ari Toubo. 1975. Contribution a l'étude de l'élevage ovin au Niger; état actuel et propositions d'amélioration. Thèse, Ecole Inter-Etats des Sciences et Médicine Vétérinaires, Dakar, Sénégal.
12. Chapter 3.4 in this book.

13.   Mason, I. L. 1980. Prolific tropical sheep. FAO Animal Production and Health
      Paper 17. Rome: FAO.
14.   Yves Berger, personal communication.
15.   Chapter 2.13 in this book.
16.   Devendra, C. 1972. Barbados Blackbelly sheep of the Caribbean. *Tropical
      Agriculture* (Trinidad) 49(1):23–29.

      Maule, J. P. 1977. Barbados Blackbelly sheep. *World Review of Animal Pro-
      duction*, 24:19–23.

      Rastogi, R. K., H. E. Williams, and F. G. Youssef. 1980. Barbados Blackbelly.
      Contributed chapter *in* (13).
17.   Lauvergne, J. J. and S. Adalsteinsson. 1976. Gènes pour la couleur de la toison
      de la brebis *corse*. *Annales de Génétique et de Sélection Animale*.
      8(2):153–172.
18.   Chapters 4.1, 4.3, 4.4, 4.5, and 4.6 in this book.
19.   Demiruren, A. S. 1980. Bahama native. Contributed chapter *in* (13).

      Katasigianis, T. S., L. L. Wilson, T. E. Cathopulius, A. A. Dorsett, Dale D.
      Fisher. 1981. Productivity of Bahama native, Florida native and Barbados
      Blackbelly sheep under improved grazing management in the Bahamas. *Tur-
      rialba* 31:113–119.
20.   Chapters 2.3, 2.4, and 2.5 in this book.
21.   Figueiredo, E.A.P. 1980. Morada Nova of Brazil. Contributed chapter *in* (13).

1.2

PRODUCTIVITY OF HAIR SHEEP AND OPPORTUNITIES FOR IMPROVEMENT

H. A. Fitzhugh, *Winrock International, U.S.A.*
G. E. Bradford, *University of California, Davis, U.S.A.*

Hair sheep are a well-adapted genetic resource contributing to the nutrition and income of producers in the tropics. However, sheep production is often secondary in importance to crops and other livestock activities. Therefore, improvements to sheep productivity must fit within the context of a broader agricultural system if they are to be accepted by producers. With this in mind, our analysis and suggestions for improving meat productivity are based on characterization of the production systems, identification of principal constraints, and evaluation of the performance traits that affect meat offtake.

PRODUCTION SYSTEMS

Hair sheep contribute to several production systems in the tropics. The amount and distribution of rainfall are often principal determinants of system characteristics, but social and economic factors are also influential. Our analysis deals with two general systems: mixed crop/livestock farming systems in medium-to-high potential agricultural areas, and livestock-based grazing systems in the drier range areas. Discussion of the role of sheep in these systems is based on information from the contributed chapters, our personal observations, and other publications (1).

Mixed Crop/Livestock Systems

Both small subsistence-level farms and large commercial operations specializing in cash crops are included in this category, with livestock usually playing a role secondary to crops. Livestock (cattle, goats, swine, poultry, and sheep) produce food for home use and also serve as a savings account to be drawn on when cash is needed.

In the traditional village systems of western Africa, sheep often roam freely within the village, subsisting on what they scavenge. Roadsides, fallow fields, and communal lands are principal grazing sites. Where sheep are a threat to growing crops, they are tethered or penned. In some instances, feed may be cut and carried to confined sheep; however, cattle will generally be given first preference for this labor-intensive activity. In the Americas, cropping and grazing pressure on available land generally precludes free roaming by sheep or other ruminants, so that tethering or penning is the rule.

Most farmers own only a few sheep, perhaps four or five. However, flocks of fifty or more are kept on large commerical farms and ranches as a source of meat for workers.

A few substantially larger commercial flocks of hair sheep have been established in Africa and the Americas. The owners have usually developed special markets for breeding stock (as in the case of Barbados Blackbelly for export) or for slaughter stock. Often sheep production is part of a larger farming operation that provides crop residues and agri-industrial by-products, such as cannery wastes, for sheep feed.

Livestock-based Systems

In pastoral, transhumant, and ranch systems, ruminants graze rangelands to produce food and income. Cattle, sheep, and goats are often managed in common herds under the care of herders. During the day, these herds may travel considerable distances in search of grazing. However, with few exceptions, they are closely confined at night as a safeguard against predators and theft.

PRODUCTION SYSTEMS CONSTRAINTS

Although the characteristics of hair sheep production systems differ from region to region, most systems are subject to the same general constraints. The three common categories of constraints are ecological, biological, and socioeconomic (2). Some examples of these constraints are listed in the following paragraphs under their respective categories. However, most production systems are affected by several, often interacting, types of constraints.

Ecological Constraints

Land (area, topography, soil fertility) and climate (rainfall, temperature, growing season) serve as constraints to sheep production, primarily determining the type and yield of plant species on which sheep feed. Of these ecological factors, only soil fertility is readily amenable to human-directed change, and then only where improvements are economically feasible.

Biological Constraints

Nutrients. We observed that there are usually restrictions in both the quantity and quality of nutrients consumed by hair sheep. For example, the common practices of limited periods of daytime grazing and overnight confinement restrict nutrient intake. Since sheep are rarely given supplemental feed, their daily nutrient intake may not be sufficient to support their full production potential. Midday heat and solar radiation further limit nutrient intake of grazing sheep. Feed supplies often vary seasonally, especially in the wet/dry tropics; feed may be abundant in the rainy season, but failure to preserve this abundance leads to dry-season deficiencies. The effects of inadequate supplies of protein and energy may be further compounded by deficiencies of essential minerals and vitamins. Lack of drinking water may be a constraint, both for sheep grazing on arid rangelands and those tethered without access to water during the day.

Health. Disease has been cited as the principal constraint to small ruminant production in the humid zone of West Africa (3). Some of the diseases listed were pleuropneumonia, trypanosomiasis, foot-and-mouth, and *peste des petits ruminants* (PPR), a rinderpest-like viral disease affecting sheep and goats. Parasites take their toll of blood and serve as vectors for disease transmission. Even when parasites can be controlled by dips, sprays, and drenches, the cost of treatment and lack of professional health services are limiting factors.

Theft and Predation. Sheep are preyed upon by both human and animal predators throughout the world. The direct cost in terms of animal loss is often overshadowed by losses in productivity that result from management practices necessary to prevent predation, such as nighttime confinement.

Genotype. For most sheep in tropical environments, genetic potential for adaptation takes precedence over genetic potential for productivity. There may even be negative genetic correlations between traits for adaptation and production. Thus, genetic improvement in production traits, such as milk yield, must generally await the resolution of nutritional and health constraints.

The vagaries of the environments in which sheep must survive favor genotypic flexibility. The same ewes that must survive through seasons (sometimes years) of drought and nutritive shortages also must be able to respond to favorable conditions with increased fertility and adequate milk production to raise healthy lambs.

Socioeconomic Constraints

Human and institutional constraints included in this category are: labor availability and management skills of producers; taste, preference, and buying power of consumers; cost and availability of credit; marketing infrastructure; and government policies on price, trade, and land tenure.

Labor and Management. Producers often choose other work or leisure activities in preference to working with sheep, because their perceived value of sheep is low.

Traditional management practices developed by trial and error through generations of experience may make efficient use of available resources, with minimal external inputs or risk to producers. However, these practices generally will not support increased productivity.

Market Demand. Sheep meat is frequently in demand to celebrate weddings and other festive occasions, but often sheep are only slaughtered for home use or sold when cash is needed. Such sporadic demand provides little stimulus for investing in management or genetic improvement.

The low value of sheep stems from lack of consumer buying power, poorly developed market infrastructure (which increases marketing costs but reduces prices received by producers), and government controls on prices. Perhaps the most visible constraints to efficient marketing are lack of efficient transportation, of a slaughter and processing infrastructure, and of market information available to producers.

PRODUCTIVITY AND EFFICIENCY

We define productivity in terms of yield of the desired product—for hair sheep, that product is principally edible tissue. (The value of by-products, such as hides and manure, might also be considered as part of the final output; however, to simplify the discussion, they are not included here.)

Efficiency of production, often expressed as the ratio of output to input, must also be taken into account. Outputs and inputs are subject not only to the cumulative effects of many traits expressed by different members of the flock (both breeding and slaughter stock), but also to many environmental factors. Thus, outputs and inputs should be considered as characteristics of the flock in the context of the total production system.

Realistic assessment of output and input may require "weighting" the component traits according to their relative importance. For example, costs of inputs and market value of outputs must be established to assess profitability or return on investment. Alternatively, energy efficiency may be assessed in terms of food energy produced relative to inputs of feed energy or fossil energy (4).

Because productivity and efficiency are such complex traits, improvements generally require a coordinated systems approach. Often benefits from improvements in one factor, such as nutrition, can be obtained only after other factors, such as health and management, also are improved. Figure 1.2.1 is a schematic representation of such relationships, indicating that the baseline productivity level of a flock under a traditional

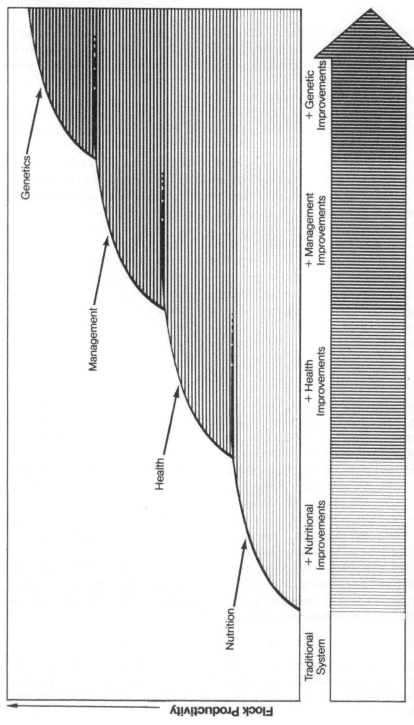

FIGURE 1.2.1 Cumulative and complementary effects of interventions to remove major constraints on flock productivity.

production system is determined by interacting environmental constraints. Generally, such flocks are producing below their genetic potential. Exceptions may occur; for example, lack of genetic resistance to disease may be the limiting factor on productivity. Usually, however, environmental constraints should merit first priority.

Interventions, such as nutritional supplementation during the dry season, can increase productivity only to a point at which other constraints become limiting. Additional benefits from nutritional improvement are then expressed as disease and parasite problems are resolved. Similarly, management practices, such as care of newborn lambs, are required to achieve still further improvements. Finally, genotypic potential may place a ceiling on further advancement in productivity. Intervention by selection and controlled mating, such as crossbreeding, can further improve the herd genotype to achieve maximum flock productivity.

Of course, Figure 1.2.1 oversimplifies the "real world" situation, because the hierarchy of constraints varies from location to location. Often disease will be the overriding constraint and must be resolved before improved nutrition can have any impact. Generally, there are complex interactions among the constraints, so that multiple interventions are required to address several environmental and genetic constraints simultaneously.

PRODUCTION TRAITS

Productivity and efficiency are the results of a composite of production traits (Table 1.2.1), with each trait subject to genetic and environmental influences. Interactions and correlations among these traits are probable, but not well documented for hair sheep.

Table 1.2.1 lists three general categories of production traits: fitness, fertility, and size-efficiency. Some traits, such as lamb survival, belong to more than one category. In Tables 1.2.2 to 1.2.11 we have summarized the means and available standard deviations for performance traits reported in the contributed chapters. Generally, these traits are the more easily measured ones such as body weight and litter size.

Other important traits were reported less frequently or did not lend themselves to summarization. Thus, we encourage readers to review the contributed chapters carefully, as they contain a wealth of information not included in this summary chapter.

Fitness

Generations of natural selection probably have reduced genetic variation for many fitness traits because unfit sheep leave few, if any, offspring. Fitness generally reflects genetic adaptation to prevailing environmental conditions.

TABLE 1.2.1. IMPORTANT TRAITS FOR SHEEP PRODUCTION IN THE TROPICS

Category	Traits
Fitness	Adaptations to environmental stress – hair coat, resistance to disease and parasites, lamb survival, longevity, temperament Adaptability to environmental fluctuation
Fertility	Prolificacy – ovulation rate, fertilization rate, embryo survival Lambing interval – postpartum interval to conception (postpartum anestrus, conception rate), gestation period Weaning rate – maternal behavior, milk production, lamb vigor Age at sexual maturity Male traits – libido, semen quality

Size and Efficiency	Growth and maturing rates
	Body weights
	    Birth weight – lamb survival
	    Slaughter weight – meat yield
	    Mature weight – maintenance requirements
	Body composition – edible tissue
	Voluntary feed intake
	Composition of diet – forage, residues, concentrates
	Efficiency of nutrient utilization for maintenance and production

The ranking of genotypes for fitness traits may differ across environments, indicating genotype × environment interactions. In some cases, general adaptability may be more important to fitness than specific adaptations. Adaptable genotypes with the flexibility to survive and produce over a wide range of conditions will be favored if environmental conditions fluctuate widely or change suddenly.

Although adaptation is difficult to measure precisely, our study suggests that hair sheep are much better adapted to tropical conditions than are wooled sheep. Contemporary fertility and lamb survival data (3) provide strong evidence supporting this advantage of hair sheep. Where imported wooled breeds or recent descendants of such breeds are grazed with hair sheep, we often saw the wooled sheep standing in the shade while the hair sheep were still grazing.

We found no clear answer to whether wool *per se* is disadvantageous in a tropical climate, or whether the difference in performance between wooled temperate breeds and hair sheep in the tropics is attributable to centuries of natural selection of hair sheep. The wooled Criollo sheep of Venezuela seem to be equal or superior to hair sheep for survival (5), and the wooled St. Elizabeth sheep of Jamaica showed less heat stress (panting) than did recently imported wool sheep. Perhaps wooled sheep can become well adapted to tropical climates, although such breeds usually develop bare bellies, necks, and tails; and their fleeces are light, of poor quality, and often matted (Photos 1.1.39 and 1.1.40). Carpet-wool breeds, such as the fat-tailed sheep of North Africa, the Middle East, and Asia, are well adapted to hot desert climates. However, these sheep are often exposed to both high and low temperature extremes, and the wool may protect against both. In hot, dry areas closer to the Equator, notably the Sahel, hair sheep are more prevalent, suggesting that wool is not a necessary protection against extreme heat or solar radiation.

Whatever the advantages or disadvantages of wool in the arid tropics, wool appears to be a very real detriment in the humid tropics. In addition to the direct effect on heat load—in an environment where heat dissipation is difficult—wool increases problems from fly strike, burrs, and external parasites. We observed that even where there are locally adapted wool sheep, hair sheep appear to be more productive and require less care in the humid tropics.

Our assessments of temperament and other behavior traits are subjective. Barbados sheep in the United States have a reputation for being nervous, with confinement requiring better-than-usual sheep fencing. Our observations of hair sheep in their "native" environments indicated that they were often more active and alert than most temperate breeds, but they did not display the nervousness associated with the Barbados sheep in the United States. However, in most cases, hair sheep are tethered, confined, or herded in close association with people. Such intensive management is almost certain to produce "well-domesticated" sheep. Certainly, the control of the herder is evident in Photo 1.1.9.

We do not know whether differences in temperament of Barbados sheep in the United States and those in Barbados (and tropical hair sheep generally) are due to the breeds with which the Barbados in the United States were crossed (such as the Mouflon), or the fact that U.S. "Barbado" sheep were feral or near-feral for several generations, or the day-to-day interaction between sheep in tropical sheep flocks and the people who care for them.

There is evidence that hair sheep are resistant to certain diseases and parasites (6). Such resistance is a major advantage, even a necessity, where prevention and treatment are not feasible because of cost and lack of health services. An example is the apparent resistance of Forest-type sheep to trypanosomiasis, which is endemic in many coastal areas of West Africa.

Fertility

The analogy that "a chain is only as strong as its weakest link" is directly applicable to the linkage of physiological events that determine fertility. These events include ovulation and conception, as well as prenatal and postnatal survival. Each event is genetically conditioned; however, each has been subjected to stringent natural selection so that additive genetic variation within populations tends to be low. Thus, environmental stresses—nutrition, disease, climate, and management—generally have the greater impact on fertility (7).

The proportion of ewes lambing, litter size, and postnatal survival are among the more easily observed traits contributing to fertility. Postnatal maternal behavior and milk production often determine whether lambs live or die. Lifetime fertility of a ewe is affected by the number of opportunities she has to go through the reproductive sequence; thus, age at sexual maturity, interval between lambings, and length of productive lifetime largely determine lifetime fertility.

Although females play the greater role in flock fertility, fertile males are essential. Traits such as libido, mating behavior, and semen quality are common measures of male fertility.

Unlike wooled sheep that have evolved in temperate regions, hair sheep show no evidence of photoperiod effect on fertility at tropical latitudes. Anestrus (when it occurs) is usually a consequence of malnutrition or undernutrition; there is also evidence of lactational anestrus. Mixed evidence for photoperiod effects was noted for hair sheep in temperate latitudes: Barbados Blackbelly ewes were fertile throughout the year at 38° N; however, ewes imported to Utah from the Virgin Islands exhibited anestrus during the winter months (8).

Size and Efficiency

These traits are linked because body size is closely related to both meat yields and nutritional requirements, which are common measures of output and input for production systems. Weights were the most commonly reported measures of size, although some linear measurements also were reported. In addition to the direct effects of body weight on individual meat yield and nutritional requirements, weights are also indirect indicators of other important production characteristics. For example, below-normal weights at birth and weaning are frequently associated with poor lamb survival in postnatal and postweaning periods, respectively.

With exceptions such as the Santa Inês breed, the hair sheep described in this book are small compared to most temperate meat-type breeds. To some extent, this small size may be due to poor nutrition, disease and parasites, and other environmental stresses. For example, ewes imported from the Virgin Islands attained adult weights of 54 kg in Utah compared to the average of 33 kg in St. Croix (9). Some of the imported sheep were from the same flocks that had the 33-kg average, suggesting that sampling effects were not necessarily the reason for the substantial difference in weight.

As long as fertility and survivability remain high, small adult size of breeding ewes may actually be an advantage. More small ewes than larger ewes can be maintained on the same quantity of nutrients; thus small adult size can be favorable with respect to flock efficiency. Also, if more animals can be kept on the same feed base, the relative cost from loss of individual sheep is lessened.

SUMMARY OF PERFORMANCE TRAITS

We calculated averages for breed types by weighting breed means from each location by the number of observations per breed at that location. For most traits, except litter size, there was usually more variation among trait means from different locations than among breed averages across locations. This is not surprising because there were substantial differences among locations in climate, nutrition, management, and other factors. Contemporary comparisons among different breeds were available at a few locations (10); however, more such comparisons are needed to support definitive statements about genetic differences.

Gestation Length

Means (Table 1.2.2) for gestation length of hair sheep are essentially the same as those usually reported for wooled sheep. Within breed groups, variation is small with standard deviations consistently below 3 days. Data from Chapter 2.1 indicate a normal distribution for gestation length, which ranged from 137 to 158 days. There is evidence of a small decrease in gestation interval due to multiple births, and gestation length for the first pregnancy may be slightly shorter than for subsequent pregnancies (11).

Lambing Interval

Lambing interval can only be measured for ewes that have two or more successive parturitions. Since there is little variation in gestation length and estrous cycle (approximately 17 to 18 days), the considerable variation in lambing interval (coefficients of variation of 12 to 32%; Table 1.2.3) is likely due to variation in length of postpartum anestrus, length of time from parturition to exposure to a ram, and rate of conception at first and subsequent estruses after parturition.

The incidence of estrus diminished markedly between January and March

TABLE 1.2.2. SUMMARY: GESTATION LENGTH, DAYS

Breed	No.	$\bar{x}$	SD	Source[a]
Pelibuey	1140	149.4	2.3	2.1.9
Pelibuey	50	148.4	2.8	2.2.1
Pelibuey	22	148.2	2.8	2.2.4
Pelibuey	70	151.8	2.3	2.3.1
West African	621	150.3	2.7	2.4.9
West African	–	148.6	–	2.5[b]
Average	1903	149.7	–	
Barbados Blackbelly	–	150.2	–	2.5[b]
Barbados Blackbelly	83	150.5	2.0	4.6.7
Morada Nova	21	149.4	1.8	2.7.4
Santa Inês	16	150.6	2.4	2.7.4
Brazilian Somali	17	148.8	1.1	2.7.7
Forest	90	148.6	2.5	3.3.3

[a] Refers to Section, Chapter, Table number.
[b] Cited in text.

for flocks in Mexico and Venezuela (12). This period coincided with the dry season in both locations; estrous activity increased markedly following onset of rains, when better nutrition would be available. We found other evidence for the impact of nutritional status on fertility traits; the percentage of ewes lambing of those exposed to rams was 75.9% for ewes on a high nutritional energy level and 64.3% for those on a low nutritional energy level.

Milk yield and suckling affected the length of postpartum anestrus. Ewes rearing twins produced more milk than those rearing singles (13). Thus, lactational stress was a probable reason why ewes rearing more than one lamb usually had longer periods of postpartum anestrus. For example, for nonlactating ewes, days from parturition to first estrus and from parturition to conception were 45.8 and 59.1; for ewes nursing one lamb, 58.6 and 74.3; and for ewes nursing two or three lambs, 85.5 and 103.0 (14). Other reports confirm this relationship between litter size and lambing interval (15). However, in one report (16), ewes that gave birth to triplets and quadruplets subsequently had shorter lambing intervals than did ewes bearing twins. A probable explanation is that mortality rate for triplets and

TABLE 1.2.3.   SUMMARY: LAMBING INTERVAL, DAYS

Breed	No.	$\bar{x}$	SD	Source
Pelibuey	370	294	75	2.1.10
Pelibuey	67	208	37	2.2.1
West African	31	214	36	2.3.1
West African	396	237	-	2.4.7
West African	1183	236	29	2.4.11
Average	2047	245	-	
Barbados Blackbelly	414	257	-	2.9.4
Barbados Blackbelly	251	253	80	2.10 [a]
Barbados Blackbelly[b]	365	253	31	2.12.1
Barbados Blackbelly	63	228	43	2.12.2
Barbados Blackbelly	200	223	42	4.6.5
Average	1293	248	-	
Forest	336	284	72	3.4.1
Yankasa	70	236	-	3.1.13
Uda	56	270	-	3.1.13

[a] Cited in text.
[b] Includes intervals between first and second parturition only.

quadruplets was substantially higher than that for twins; thus, of the ewes giving birth to more than two lambs, many nursed only singles or, perhaps, no lambs at all.

Average lambing intervals were approximately 8 months with no appreciable difference among breeds. Gestation length was approximately 5 months and the average postpartum interval to conception was approximately 3 months (or about the age when most lambs were weaned). Thus, a program of three lambings every 2 years should not necessarily affect lamb survival and postnatal growth rate. Some experiments support the feasibility of three lambings in 2 years (12); however, improved nutrition and management of the breeding flock are usually necessary.

Litter Size

More data were available for this important and easily observable trait than for other traits (Tables 1.2.4 and 1.2.5). Breed means suggest substantial differences among breeds, both for average and distribution of litter size. Average litter sizes for American breeds were Barbados Blackbelly, 1.84; Virgin Islands, 1.61; Pelibuey–West African, 1.24; and Blackhead Persian, 1.08.

TABLE 1.2.4. SUMMARY: LITTER SIZE

Breed	No. of Parturitions	Litter Size		Source
		$\bar{x}$	SD	
Pelibuey	1781	1.22	.46	2.1.13
Pelibuey	128	1.20	-	2.2.1
West African	70	1.29	.54	2.3.1
West African	287	1.14	.36	2.4.10
West African	277	1.43	.53	2.5.2
West African	130	1.26	.38	2.8.3
Average	2673	1.24	-	
Virgin Islands	73	1.54	-	2.11.2
Virgin Islands	18	1.44	-	4.2.5
Virgin Islands	38	1.82	-	4.2.6
Average	129	1.61	-	
Barbados Blackbelly	130	1.71	.75	2.1.13
Barbados Blackbelly	195	1.45	.60	2.5.2
Barbados Blackbelly	41	1.66	.58	2.6.1
Barbados Blackbelly	68	1.35	.54	2.8.3
Barbados Blackbelly	362	2.10	-	2.9.5
Barbados Blackbelly	1031	1.99	.78	2.9.6
Barbados Blackbelly	354	2.02	-	2.9.8
Barbados Blackbelly	489	1.86	.78	2.10.1
Barbados Blackbelly	365	1.39	.54	2.12.1
Barbados Blackbelly	81	1.98	.81	2.12.2
Barbados Blackbelly	95	1.65	.52	4.1.6
Barbados Blackbelly	31	1.42	-	4.2.6
Barbados Blackbelly	66	1.70	-	4.5.1
Barbados Blackbelly	272	1.81	.63	4.6.1
Average	3580	1.84	-	
Santa Inês	16	1.25	.45	2.7.4
Morada Nova	21	1.76	.54	2.7.4
Blackhead Persian	83	1.04	.19	2.5.2
Blackhead Persian	38	1.10	.31	2.8.3
Brazilian Somali	17	1.23	.44	2.7.7
Average	138	1.08	-	
Forest	381	1.46	-	3.1.4
Forest	683	1.11	-	3.3.5
Forest	169	1.12	.35	3.4.1
Average	1233	1.22	-	
Peul	65	1.02	-	3.4.2
Peul	282	1.06	.24	3.5.4
Touabire	28	1.11	-	3.4.2
Yankasa	102	1.25	-	3.1.13
Yankasa	31	1.26	-	3.1.15
Uda	71	1.14	-	3.1.13
Uda	25	1.44	-	3.1.15
Average	604	1.12	-	

TABLE 1.2.5. SUMMARY: DISTRIBUTION OF LITTER SIZE

Breed	No. of Parturitions	Litter size, % of parturitions				Source
		One	Two	Three	Four	
Pelibuey	1781	80.4	17.6	2.0	0	2.1.13
West African	287	86.4	13.2	0.3	0	2.4.10
West African	277	58.8	39.4	1.8	0	2.5.2
West African	130	73.8	26.2	0	0	2.8.3
Average	2475	78.3	20.0	1.7		
Barbados Blackbelly	130	43.8	44.6	8.5	3.1	2.1.13
Barbados Blackbelly	195	61.0	33.3	5.6	0	2.5.2
Barbados Blackbelly	41	39.0	56.1	4.9	0	2.6.1
Barbados Blackbelly	68	67.6	29.4	3.0	0	2.8.3
Barbados Blackbelly	1031	28.4	47.5	21.2	2.8[a]	2.9.6
Barbados Blackbelly	489	34.8	46.9	15.8	2.5[b]	2.10.1
Barbados Blackbelly	365	63.3	34.0	2.7	0	2.12.1
Barbados Blackbelly	95	36.8	61.1	2.2	0	4.1.6
Barbados Blackbelly	272	30.9	57.7	11.0	0.3	4.6.1
Average	2686	39.2	45.6	13.5	1.7	
Santa Inês	16	75.0	25.0	0	0	2.7.4
Morada Nova	21	28.6	66.7	4.8	0	2.7.4
Blackhead Persian	38	89.5	10.5	0	0	2.8.3
Brazilian Somali	17	76.5	23.5	0	0	2.7.7
Peul	282	94.7	5.0	0.3	0	3.5.4

[a] Includes one litter of quintuplets.
[b] Includes two litters of quintuplets.

There were few observations for breeds in Brazil; however, the Morada Nova appears to be more prolific than the Santa Inês or Brazilian Somali. The Brazilian Somali had a larger average litter size than did the similar fat-rumped Blackhead Persian; however, the appearance (Photo 1.1.32) of Brazilian Somali sheep suggests that substantial outcrossing to nonfat rumped breeds has occurred.

The African Forest type had an average litter size of 1.22, similar to that of the American Pelibuey–West African. The average for the Sahelian type was 1.12; however, there was considerable variation among the means combined in this average. Multiple births and the resulting lighter, perhaps weaker, lambs may be a major disadvantage in arid regions where flocks must travel considerable distances for feed and water.

Ewes lambing for the first time generally had fewer multiple births than did previously parous ewes (17).

The size of the litter appears to be a moderately repeatable trait (18). Correlations for size of adjacent litters are less than correlations for size of nonadjacent litters, suggesting that the effect of litter size on ewe condition during breeding season may influence the size of the next litter. There is some evidence of seasonal effects on litter size; ewes conceiving in the dry season had smaller litters than did those conceiving in the rainy season (19).

The distribution pattern for litter size differs among breeds (Table 1.2.5). The most frequently occurring litter size was twins for Barbados Blackbelly and Morada Nova, with singles most common for all other breeds. Barbados Blackbelly ewes also frequently produced triplets, with quadruplets not uncommon.

Among all hair sheep breeds, Barbados Blackbelly ewes had the highest average litter size with means of 2.0 achieved under good conditions. Variability in litter size of Barbados Blackbelly also appears to be greater than for other breeds. Even in the flocks with lower mean values, the coefficient of variation was high in absolute terms and in comparison with values for other breeds at the same location (20).

We are not sure why there is greater variability for the Barbados Blackbelly breed; however, the possibilty of a major gene affecting variability in litter size cannot be dismissed. This hypothesis is being tested in the Booroola Merino, another strain of sheep with an unexpectedly high mean and coefficient of variation in litter size (21). An increase in variability in litter size could contribute to larger average litter sizes in an environment where sheep are managed in small flocks and where feed supply is adequate all or most of the year (i.e., where survival of twins, triplets, and quadruplets is sufficiently high so that multiple births mean more progeny that will reach breeding age). A major gene could be maintained in the population by selection and over time a breed characterized by high mean litter size and above-normal variability in litter size could evolve. This is a hypothesis worth testing in crosses with the Barbados Blackbelly. Data on crosses with Barbados Blackbelly provide evidence of this breed's ability to transmit its high potential for prolificacy (22).

Lamb Mortality

Postnatal death losses were substantial; in a majority of the flocks, average losses exceeded 25% (Table 1.2.6). Many factors contributed to these losses—lamb vigor, litter size, maternal behavior, milk production, and disease challenge.

Some of the greatest losses occurred in the period immediately after birth. In one flock, two-thirds of the deaths of Barbados Blackbelly lambs occurred in the first week after birth (23). Poor lamb vigor (often associated with low birth weights) and lack of maternal attention contributed to these early losses. Lambs that died in their first week had an average birth weight of 1.7 kg, compared to the flock average of 2.5 kg. Data from another flock (24) further support this relationship between birth weight and first-week mortality rates: the mortality rate was 68% for lambs weighing less than 1

TABLE 1.2.6. SUMMARY: LAMB MORTALITY, %

Breed	Interval	No. Births	% Mortality	Source
Pelibuey	Birth-17 weeks	453	11.2	2.1[a]
West African	Birth- 8 months	731	24.6	2.4.12
West African	Birth- 8 months	64	23.4	2.4.14
West African	Birth- 6 months	396	25.7	2.5.13
Average[b]		1644	21.1	
Virgin Islands	Birth-weaning	113	11.5	2.11.2
Barbados Blackbelly	Birth-17 weeks	222	45.8	2.1[a]
Barbados Blackbelly	Birth- 6 months	282	34.5	2.5.13
Barbados Blackbelly	Birth- 4 months	68	43.0	2.6.1
Barbados Blackbelly	Birth- 3 months	612	21.9	2.9.7
Barbados Blackbelly	Birth- 1 month	1026	10.9	2.10.6
Average[b]		2210	21.5	
Forest	Birth- 4 months	755	23.4	3.3.5
Forest	Birth- 6 months	190	47.9	3.4.1
Average[b]		945	28.3	
Blackhead Persian	Birth- 6 months	86	34.6	2.5.13

[a] Cited in text.
[b] Averages include postnatal periods of different lengths.

kg; 30% for lambs weighing 1 to 1.5 kg; 5% for lambs weighing 1.5 to 2 kg; and 4% or less for lambs weighing more than 2 kg.

Mortality rates generally were higher for lambs born in litters of three or more as compared to rates for singles and twins (Table 1.2.7). Lambs born in larger litters were generally lighter in weight (Table 1.2.8) and probably less vigorous. Ewes may not have given as much individual attention to lambs in large litters. An obvious problem is the inability of ewes to suckle more than two lambs at a time; thus, all lambs in large litters are unlikely to receive equal and adequate shares of colostrum and milk. Although cross-nursing may occur (Photo 1.2.1), the larger, more vigorous lambs are likely to benefit most.

Several contributed reports documented the critical importance of the ewe's nutritional and physiological status to lamb survival, generally through the effect on milk yield. For example, improving nutrition of ewes during pregnancy reduced lamb mortality from 23 to 11% (24). Mortality

TABLE 1.2.7. EFFECT OF LITTER SIZE ON LAMB MORTALITY RATE, %[a]

Interval	Litter size					Source
	1	2	3	4	5	
Birth - 17 weeks	15	50	75	53	-	2.1
Birth - 3 months	18	16	32	23	-	2.9.7
Birth - 1 month	5	6	20	32	40	2.10.6

[a] Barbados Blackbelly flocks.

TABLE 1.2.8. SUMMARY: BIRTH WEIGHT, kg

Breed	Singles		Twins		Triplets		Quadruplets		Source
	No.	$\bar{x}$	No.	$\bar{x}$	No.	$\bar{x}$	No.	$\bar{x}$	
Pelibuey	1432	2.7	344	2.1	36	1.7	-	-	2.1.16
Pelibuey	88	2.7	33	2.1	-	-	-	-	2.2.2
West African	50	2.6	40	2.2	-	-	-	-	2.3.1
West African	366	2.6	248	2.1	48	1.7	16	1.4	2.4.15
West African	163	3.1	218	2.6	15	1.9	-	-	2.5.4
Average	2099	2.7	883	2.2	99	1.7	16	1.4	
Virgin Islands	27	3.1	81	2.7	30	2.7	8	2.1	4.2.1
Barbados Blackbelly	57	2.7	116	2.0	34	1.7	14	1.8	2.1.17
Barbados Blackbelly	119	2.7	130	2.5	33	2.0	-	-	2.5.4
Barbados Blackbelly	17	2.6	46	2.2	6	2.2	-	-	2.6.2
Barbados Blackbelly	173	3.2	555	3.1	380	2.4	49	2.1	2.9.2
Barbados Blackbelly	49	3.5	148	2.9	70	2.7	12	2.5	2.10.5
Average	415	3.0	995	2.8	523	2.4	75	2.1	
Forest	202	1.9	310	1.6	43	1.4	-	-	3.1.7
Forest	488	1.7	122	1.5	-	-	-	-	3.3.7
Average	690	1.8	432	1.6	43	1.4	-	-	
Santa Inês	12	3.3	6	2.5	-	-	-	-	2.7.12
Morada Nova	6	3.0	28	2.3	-	-	-	-	2.7.12
Blackhead Persian	80	2.5	6	2.1	-	-	-	-	2.5.4
Brazilian Somali	13	2.4	8	1.7	-	-	-	-	2.7.13
Average	93	2.5	14	1.9	-	-	-	-	
Peul	80	3.1	28	2.7	3	2.8	-	-	3.5.5

1.2.1 Example of crossnursing with multiple, unrelated lambs attempting to suckle Pelibuey ewe, Mexico.

of lambs born to primiparous ewes (1 to 2 years old) was 37%, compared to 18% for mature ewes (24). Under African village conditions (25), 22% of ewes lambing in intervals of less than 7 months aborted or produced stillborn lambs, compared to no abortion or stillbirths for ewes lambing in intervals greater than 7 months.

These data show that surviving the first week after birth was no guarantee of a lamb's survival. Poor nutrition, disease, and parasites often led to high death losses of lambs in the period before and, especially, after weaning. For example, in a flock of Forest sheep in Senegal (26), mortality from birth to 10 days (including stillbirths) was only 4.8%, but 45.6% of the lambs alive at 10 days of age died before they reached 6 months of age.

Preweaning survival rates were generally higher for female than for male lambs (27).

Across-location averages for mortality of Barbados Blackbelly and Pelibuey lambs were essentially equal (Table 1.2.6). However, contemporary comparisons of these breeds indicate mortality rates are substantially higher for Barbados Blackbelly (28). The effects of differences in litter size and birth weight are confounded in these breed comparisons.

Although the high death losses reported for hair sheep are a cause for concern, such losses are higher for lambs from wooled breeds imported to the tropics (29). In a report for a temperate environment, lamb survival of hair sheep crosses was superior to that of lambs from contemporary wooled breeds and crosses (30).

Body Weights

Although some linear measurements are presented in the contributed chapters, body weights were the most commonly reported measures of size. Weights for lambs at birth and weaning and for ewes are summarized in Tables 1.2.8 to 1.2.11. Standard deviations for some of these weights are available in the source tables.

Birth Weight. Breed averages for birth weights of single lambs ranged from 3.3 kg for Santa Inês to 1.8 kg for Forest types (Table 1.2.8). Averages for male lambs were approximately 3 to 5% heavier than those for female lambs. Breed averages for birth weight of single lambs as a percentage of breed averages for adult ewe weight were 7 to 9%; the higher percentages were for the Blackhead Persian and Sahelian breeds, which also were the least prolific breeds.

Average birth weights of twins, triplets, and quadruplets were progressively lighter than those of singles, as expected. However, there were differences among breeds in this relationship between litter size and birth weight. Average weights for twins, triplets, and quadruplets as percentages of weight of singles were: for Pelibuey—twins 81%, triplets 63%, and quadruplets 52%; for Virgin Islands sheep—87, 87, and 68%; and for Barbados Blackbelly—93, 80, and 70%. As discussed in Chapter 2.10, the pattern of relatively heavy birth weights from large litters of Virgin Islands and Barbados Blackbelly lambs is similar to the pattern for Finnish Landrace, another "prolific" breed. Since birth weight is related to lamb survival, this pattern may be an important factor in the successful evolution of prolificacy. However, the probable increase in physical and physiological stress on ewes carrying these larger, heavier litters could adversely affect their longevity and certainly suggests need for better nutrition of these ewes during pregnancy. We have already noted the advantage of improved nutrition for pregnant ewes on postweaning lamb survival.

Weaning Weight. Breed averages for weaning weights were remarkably

TABLE 1.2.9. SUMMARY: WEANING WEIGHT, kg

Breed	Weaning age, days	Litter size	Male No.	Male $\bar{x}$	Male Ratio[a]	Female No.	Female $\bar{x}$	Female Ratio[a]	Source
Pelibuey	120	Single	54	12.7		47	12.2		2.1.19
		Multiple	17	10.6	83	18	10.1	83	
West African	120	Single	28	16.9		60	13.9		2.4.16
		Twin	7	11.6	69	20	10.9	78	
		Triplet	3	9.0	53	–	–		
Barbados Blackbelly	120	Single	19	12.4		27	10.6		2.1.20
		Twin	28	11.3	91	26	10.4	98	
		Triplet	2	15.8	127	7	10.6	100	
		Quad	3	10.6	85	4	8.2	77	
Barbados Blackbelly	120	Single	3	13.7		6	13.8		2.6.2
		Twin	10	13.4	98	14	12.3	89	
Barbados Blackbelly	75[b]	Single	34	12.0		45	11.6		2.9.3
		Twin	109	9.7	81	105	9.3	80	
		Triplet	82	8.8	73	62	8.1	70	
		Quad	6	8.1	67	4	7.1	61	
Morada Nova	210	Single	3	14.0		3	13.1		2.7.12
		Twin	7	9.8	70	9	10.5	80	
Santa Inês	210	Single	5	15.0		4	11.7		2.7.12
		Twin	–	–	–	2	13.4	114	
Brazilian Somali	95	Single	4	12.9		9	12.7		2.7.13
		Twin	2	9.2	71	6	8.2	64	

[a] Ratio of weights for multiple births to single births.
[b] Approximate average age at weaning.

similar across locations, especially considering the differences in weaning age and preweaning environment (Table 1.2.9). Standard deviations for weaning weights were approximately 2 to 3 kg, with coefficients of variation of approximately 20%. Multiple-birth lambs generally weighed less at weaning age than single-born lambs, reflecting the effects of competition for milk and carryover effects of weight at birth.

Postweaning Gain. Daily gains of 170 g were achieved by West African lambs under semiconfinement conditions (24), and daily gains of 260 g were recorded for well-fed Virgin Islands lambs in temperate conditions (31). However, postweaning gains of lambs grazing native tropical pastures rarely exceeded 100 g and were often as low as 50 g per day (32). Assuming a 100-day weaning weight of 12 kg, and a target slaughter weight of 25 kg, lambs gaining 100 g per day would be 230 days old at slaughter, but a rate of gain of only 50 g per day would mean a 360-day slaughter age.

TABLE 1.2.10. SUMMARY: AGE AND WEIGHT AT PUBERTY AND FIRST PARTURITION

Breed	Postweaning treatment		Puberty			First lambing		
		No.	Age, days	Wt, kg	No.	Age, days	Wt, kg	Source
Pelibuey	Confined, concentrate	92	321	23	61	480	26	2.1.4
	Grazing, 24-hours	105	404	22	56	555	26	and
	Grazing, 8-hours	63	429	22	48	573	25	2.1.5
West African	(1) Pasture	196	286	21	-	-	-	2.4.4
	(2) Pasture	20	305	21	-	-	-	2.4.4
	Pasture, supplement	19	269	22	-	-	-	
	(3) Single born	98	262	-	93	418	-	2.4.5
	Multiple born	72	312	-	69	463	-	
Morada Nova	Grazing, 7-hours	6	214	21	6	498[a]	29	2.7.5
Santa Inês	Grazing, 7-hours	6	220	29	5	450[a]	39	2.7.5
Brazilian Somali	Grazing, 7-hours	18	284	20	3	490[a]	28	2.7.5
Forest	Grazing, supplement	21	259	16	-	-	-	3.3[b]

[a] Average age at first mating was: Morada Nova, 350 days; Santa Inês, 287; Brazilian Somali, 342.

[b] Cited in text.

Weight and Age at Sexual Maturity. Ewe lambs tended to reach puberty at approximately 60 to 65% of adult weight. Thus, postweaning environment—through its effect on rate of postweaning gain—was a major determinant of age at sexual maturity and, more importantly, of age at first parturition (Table 1.2.10). Within-location comparisons of effects of postweaning treatments illustrate this point (12). Improving nutrition—whether by supplementation or by longer grazing periods—reduced age at sexual maturity but had little effect on weight. There also seems to be a carryover effect of type of birth on maturing rate, because multiple-birth lambs were 50 days older at puberty than were single-birth ewe lambs.

Other data on age and weight at sexual maturity and/or first lambing are presented in contributed chapters (13). Among Nigerian Forest ewes, 37% of the ewes lambed before 12 months and 66% before 15 months of age. Pelibuey ewes were 245 days old and weighed 23 kg at puberty. Barbados Blackbelly ewes, well fed in drylot, began to cycle at approximately 5 months of age, and the average age at first lambing was 370 days for 64 ewes.

Adult Weight. Weights of adult ewes may vary considerably during the year, depending on nutritional status, stage of pregnancy, and milk production. Most of this variation after the ewe attains full growth of lean meat

TABLE 1.2.11. SUMMARY: ADULT EWE WEIGHT, kg[a]

Breed	No.	$\bar{x}$	SD	Source
Pelibuey	174	37	-	2.1.1
Pelibuey	358	31	4.4	2.1.11
Pelibuey	55	33	4.8	2.2.1
Pelibuey	15	37	2.9	2.3.2
West African	92	39	5.8	2.4.1
West African	64	30	-	2.8.2
Average	743	34	-	
Virgin Islands	134	33	6.1	2.11.1
Virgin Islands	12	54	12.4	4.2.1
Virgin Islands	22	37	-	4.2.2
Average	168	35	-	
Barbados Blackbelly	33[b]	33	6.3	2.6.1
Barbados Blackbelly	98[c]	32	-	2.8.2
Barbados Blackbelly	172	44	-	2.9.4
Average	270	40	-	
Blackhead Persian	11	36	2.7	2.3.2
Blackhead Persian	29	28	-	2.8.2[d]
Brazilian Somali	17	26	-	2.7[d]
Average	46	27	-	
Forest	164	29	-	3.1.9
Forest	97	25	-	3.4.1
Average	261	27	-	
Morada Nova	150	30	3.9	2.7.8
Peul	154	31	3.6	3.4.2[d]
Peul	48	35	4.9	3.5[d]
Touabire	65	35	4.4	3.4.2
Average	267	33	-	

[a] Averages are usually for ewes 2 years old and older.
[b] Two-year-old ewes.
[c] Three-year-old ewes.
[d] Cited in text.

and bone is due to variation in body fat. Fat can account for 20% or more of differences in body weight. For example, 154 Peul ewes, observed in very thin condition, probably would have averaged 6 to 7 kg heavier when in good condition (33). We have previously mentioned the possible effects of nutrition, health status, and other environmental factors on weights of Virgin Islands ewes in St. Croix compared to those in Utah.

Although little information was available about degree of fatness, the breed means in Table 1.2.11 are generally indicative of commonly perceived size differences among breeds. Excepting the Santa Inês breed for which ewe weights of 60 to 70 kg were cited (34), the Barbados Blackbelly is generally considered to be the largest American hair sheep and the Blackhead Persian the smallest. Among the breeds in West Africa, the Sahelian types are definitely larger than the Forest types.

A few examples of ram weights are given in the contributed chapters. Rams kept for breeding are a selected set, probably heavier than the population average. Reported weights for Pelibuey–West African rams included 6 rams averaging 49 kg (23); 3 rams averaging 49 kg, with one ram weighing 60 kg (35); and 11 rams averaging 57 kg (24). Average height at withers of these rams ranged from 65 to 70 cm. Weights cited for rams of Brazilian breeds (34) included: Santa Inês, averaging 80 kg, with some rams exceeding 100 kg; Morada Nova, averaging 39 kg; and Brazilian Somali, averaging 40 to 60 kg. Weights of rams of the Sahelian breed, Peul, included 10 rams averaging 48 kg (26).

FLOCK PRODUCTIVITY AND EFFICIENCY

The emphasis of our discussion to this point has been on specific performance traits expressed by individual sheep; however, emphasis is more appropriately placed on the cumulative effects of the many traits that determine lifetime productivity. Although individual sheep conceive, grow, live and die, the performance of individuals has primary relevance in terms of the net impact on productivity and efficiency of the flock as a whole. The flock, not the individual, is the production unit of concern.

With these considerations in mind, we have developed a Flock Productivity Index (FPI) and a Flock Efficiency Index (FEI) incorporating the many traits that determine flock productivity and efficiency. When these indices are used in comparisons among flocks (or among breeds), the findings are more meaningful than are comparisons of individual traits. For example, some important traits such as litter size and lamb mortality may be negatively correlated and flocks might rank differently for each trait. In such cases we have little basis for determining which flock or breed is "best." However, rankings for total productivity or efficiency can identify the "best."

Description of Indices

Traits included in FPI are fertility (F), lamb survivability (S), ewe longevity (L), and meat yield (Y); i.e., FPI = F × S × L × Y, where

Fertility	=	lambs / ewe / day
	=	(lambing rate × prolificacy) / lambing interval

and

Lambing rate	=	ewes lambing / ewes exposed
Prolificacy	=	lambs born / ewes lambing
Lambing interval (in days)	=	average time between consecutive lambings
Survivability	=	lambs slaughtered / lambs born
Longevity (in days)	=	average culling age − average age at first lambing
Yield (kg / day)	=	edible portion of carcass / slaughter age

All of these traits may not be available for all flocks. Sometimes the missing values may be estimated. For example, edible portion can be estimated as approximately 70% of carcass weight, which in turn is approximately 40% of slaughter weight. However, when a trait is missing for all or even most flocks, a suitable proxy trait may be substituted. Average weight per day of age at slaughter is a useful proxy for edible portion per day of age. If slaughter weight is not available for all flocks, weight at some other age for which comparable data are available may be used.

The Flock Efficiency Index estimates output/input. Types of inputs that might be used to evaluate efficiency include feed energy requirements, hectares of land, hours of labor, and capital requirements.

Again, use of proxies for inputs may be necessary. For example, feed energy requirements for the flock are a principal but difficult-to-measure input. However, because the breeding flock utilizes approximately 70% of the total feed energy (3), a useful proxy for flock requirements is the average physiological weight of adult ewes ($W^{.75}$). Thus, an appropriate measure of flock efficiency is FEI = FPI / $W^{.75}$.

Refinements to these indices can improve their precision and accuracy, assuming needed information is available. One type of index, the selection index, incorporates economic values, heritabilities, and correlations among traits to provide the criterion for selection among individuals (36). Similar refinements could be incorporated in indices developed for comparisons of flocks, breeds, and other groups.

Comparisons Among Breeds

We have applied the general concepts discussed in the previous section to comparisons among breeds of hair sheep, using the across-location averages presented earlier in this chapter. Since averages for all traits were not available for all breeds, we used simplified indices FPI* and FEI* and substituted proxies for missing values.

$$FPI^* = \text{(litter size} \times \text{lamb survival} \times \text{birth weight)} / \text{lambing interval}$$
$$FEI^* = FPI^* / \text{(adult ewe wt)}^{.75}$$

These modified indices have many limitations. No account is taken of lambing rate (ewes lambing / ewes exposed) or ewe longevity, which are important components of flock productivity. The averages for lamb survival derived from average mortality (Table 1.2.6) involve different postnatal time intervals. Average birth weight has limited value as an indicator of weight per age at slaughter; weaning weight would have been a better choice. However the data for weaning weights (Table 1.2.9) were not as complete as those for birth weights (Table 1.2.8). Using estimates for missing breed averages introduces errors of unknown magnitude and reduces the variation between breeds. Nevertheless, these indexes do summarize the effects of many traits on productivity and efficiency and serve as examples of the use of indices.

Rankings for FPI* and FEI* were: Barbados Blackbelly, Virgin Islands, Pelibuey–West African, Blackhead Persian, Forest (Table 1.2.12). The appreciable advantage for Barbados Blackbelly in FPI* was reduced for FEI*, especially in comparison to the Virgin Islands sheep. The low ranking of the Forest sheep for FPI*, even in comparison to the less prolific Blackhead Persian, resulted from their longer lambing interval and lighter birth weight. We could not determine from available data the extent of environmental effects on breed averages. For example, the lower ranking of Forest sheep may have been partially due to poorer environmental conditions in Africa as compared to those in the Americas. Contemporary comparisons at the same locations would provide more suitable data for ranking breeds.

The units for analysis for FPI* are grams of lamb weight produced per day of lambing interval. Units for FEI* are grams of lamb weight per kg metabolic weight of adult ewes per day of lambing interval. FPI* and FEI* can be put on an annual basis by multiplying by the average number of

Table 1.2.12. BREED AVERAGES FOR PRODUCTION TRAITS AND INDICES

Trait	Pelibuey, West African	Virgin Islands	Barbados Blackbelly	Blackhead Persian	Forest
Litter size, no. lambs	1.24	1.61	1.84	1.08	1.22
Lambing interval, days	245	248[a]	248	248[a]	284
Lamb survival	0.79	0.78[a]	0.78	0.65	0.72
Birth weight, kg	2.5	2.7	2.7	2.4	1.7
Ewe weight, kg	34	35	40	27	27
$W^{.75}$	14.1	14.4	15.9	11.8	11.8
FPI* x 10^3	10.0	13.6	15.6	6.8	5.3
FEI* x 10^3	0.71	0.95	0.98	0.58	0.45

[a] Average for Barbados Blackbelly substituted for unknown value.

lambings expected per year (i.e., 365 days/days in lambing interval). This multiple would be approximately 1.5 for all breeds except the African Forest type, for which only 1.3 lambings would be expected.

If birth weight is approximately 10% of slaughter weight, multiplying the indices by 10 would convert to a slaughter weight basis. Based on post-weaning gains of 100 and 50 g per day, age at slaughter would be approximately 230 and 360 days, respectively (see discussion earlier in this chapter); thus, an average slaughter age of 295 days could be used. Lambing rates of 80 to 90% were reported for Pelibuey ewes (23). Age at sexual maturity ranged from 7 to 14 months (Table 1.2.10), and ewes might be expected to remain in the breeding flock until 60 to 70 months of age, which means a reproductive longevity of approximately 50 months or 1500 days.

Using these additional estimates of average values for important production traits, FPI and FEI could be derived from FPI* and FEI*. However, if the same values were used for all breeds (as they would be in the absence of any evidence for different values for different breeds), ranking for the derived FPI and FEI would remain the same as for FPI* and FEI*.

These examples indicate the critical need for collecting more comprehensive data on each of these breeds, preferably from contemporary comparisons in the same environment.

Adequate characterization of hair sheep productivity on the basis of life cycles depends on accurate measurement of many traits. A minimal set for experimental observation would include:

- Classification traits—breeding, sex, age
- Fitness traits—coat type, date and cause of death
- Fertility traits—parturition dates, litter size

1.1.7 Santa Inês ram, Brazil.

1.1.8 Barbados Blackbelly
ewe with quintuplets,
Barbados.

1.1.9 Sahelian sheep in
Bamako market, Mali.

1.1.10 Santa Inês ewe, Brazil.

1.1.11 Sahelian-type ewe, Mali.

1.1.12 Virgin Island white ram, St. Croix.

1.1.13 Polled Forest-type ram with heavy mane and throat ruff, Ivory Coast.

1.1.14 Barbados Blackbelly ewes, Mexico.

1.1.15 Forest-type ewe, blackbelly pattern, Cameroon.

Comparisons of Hair Sheep in the Americas and Western Africa

Sahelian-type rams with
spiral horn, pendulous ears,
no mane, and long tails:
1.1.16 Nigeria (top left);
1.1.17 Mali (top right);
1.1.18 Bouaké market,
Ivory Coast (left).

Forest–Savannah-type rams
with crescent horns, short
ears, manes, and relatively
short tails:
1.1.19 Ivory Coast (right);
1.1.20 Liberia (bottom right).

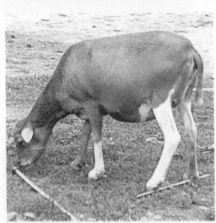

Forest–Savannah-type ewes in West Africa showing variety of coat color and patterns.
Clockwise from top left: 1.1.21 Liberia; 1.1.22 Ivory Coast; 1.1.23 Ivory Coast;
1.1.24 Cameroon; 1.1.25 Cameroon; 1.1.26 Cameroon.

Hair sheep rams in the Americas. Photos clockwise from top left: 1.1.27 Barbados Blackbelly, Mexico; 1.1.28 Pelibuey, Mexico; 1.1.29 Africana, Venezuela; 1.1.30 Santa Inês, Brazil; 1.1.31 Morado Nova, Brazil; 1.1.32 Brazilian Somali, Brazil.

Hair sheep ewes in the Americas. Photos clockwise from top left: 1.1.33 Virgin Island ewe with twins; 1.1.34 Africana, Colombia; 1.1.35 Pelibuey ewes, lamb in foreground showing payaso pattern; 1.1.36 Blackbelly pattern, Jamaica; 1.1.37 Blackhead Persian fat-rumped ewes, Colombia; 1.1.38 Santa Inês ewe, probably a crossbred, Brazil.

1.1.39 Coarse Wool Italian breed, Bergamasca, and Pelo do Boi, Brazil.

1.1.40 Crosses of Barbados Blackbelly with wooled breeds, Jamaica.

1.1.41 (left) Poor quality, matted fleeces on crosses. From left to right: predominantly Africana, Africana × Criollo, and Criollo sheep in Venezuela.

1.1.42 (below) Wiltshire Horn ram, Tortola.

1.1.43 (below) Katahdin ewe with hairy lamb. Katahdin was developed in the United States from crosses of Virgin Island sheep with Suffolk and Wiltshire Horn breeds.

1.1.44 Barbados Blackbelly flock in Mexico.

1.1.45 Barbados Blackbelly
flock in Guyana.

1.1.46 "Barbado" rams
developed in Texas from
crosses of Barbados Blackbelly, Rambouillet, and
Mouflon. They sometimes are
hunted as game animals.

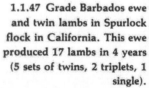

1.1.47 Grade Barbados ewe
and twin lambs in Spurlock
flock in California. This ewe
produced 17 lambs in 4 years
(5 sets of twins, 2 triplets, 1
single).

- Size traits—weight at (and date of) birth, weaning, slaughter, and parturition.

Other traits could be added but this minimal set would provide information needed to assess productivity and efficiency. As results from different locations are reported they can be pooled to develop a general overview of hair sheep performance.

OPPORTUNITIES FOR RESEARCH AND DEVELOPMENT

Few populations of hair sheep are producing to their full potential. One reason is that relatively little effort has been devoted to improving hair sheep and their production environment. We feel that there is interesting, important applied research to be done with hair sheep. Our experiences suggest that such research should be problem-oriented, designed to resolve major constraints, and that, by contrast, most basic research with sheep should remain in the hands of scientists and institutions in developed countries. Thus the emphasis will be on applied research in developing countries, with more basic research to be done by scientists and institutions in developed countries. This is because there are so many top-priority problems of an applied nature in developing countries, not because of a lack of qualified scientists. For example, scientific resources required for basic research on the nutritional requirements of hair sheep would be better utilized in testing the feasibility of low-cost rations based on locally available feedstuffs. Nutritional requirements determined for wooled sheep in temperate environments can serve as first-order approximations of the requirements for hair sheep in tropical environments.

Recognizing that specific priorities and procedures for research and development will vary for different systems, our recommendations concentrate on general opportunities observed during the course of this study.

The first priority for research should be characterization of hair sheep resources and their production environments. This research should identify the principal constraints on productivity and provide the necessary understanding of traditional systems. The characterization should entail year-round monitoring of feed supply, health status, and performance levels of sheep, plus other aspects of the production system. An important benefit of the characterization phase is that the researcher will become thoroughly familiar with the traditional system and thus be better able to design research relevant to the needs of producers.

Examples of research opportunities that are likely to be identified by this characterization phase include:

- Utilization of low-cost feedstuffs, including food and cash crop residues and forage from fallow or marginal lands.
- Procedures for preserving nutritional quality of surplus feeds for use in periods of scarcity. Hay- and silage-making techniques that may work well in temperate climates are often inappropriate to the tropics.
- Flock health programs suited to needs of producers in developing countries. Costs must be kept low because the value of sheep is often low and financial resources of producers are limited.
- Management procedures to improve reproductive efficiency— including comparison of continuous exposure to rams with limited mating periods in which timing is coordinated with availability of feed supplies or other resources.
- Procedures for improving preweaning and postweaning survival of lambs through improved nutrition and management, to reduce this substantial wastage of animal resources.
- Establishment of appropriate goals and strategies for genetic improvement. Research should include evaluation of genetic and environmental sources of variation in performance traits, and comparison of breeds and crosses for productivity and efficiency.

The results of this research combined with practical experience can lead to effective development programs for utilizing the full potential of hair sheep. It is important that interventions be evaluated under actual farm conditions before they are widely promoted. On-farm evaluation will help insure that interventions are appropriate to the needs and resources of farmers.

At the producer level, these interventions will probably involve development of technology packages to improve feed production and preservation, flock health, and management. Genetic improvements will generally require government-supported breeding and multiplication centers that maintain populations of hair sheep of sufficient numbers to permit effective selection and breeding for improved genotypes. Trained personnel will be needed to deliver these technology packages and improved stocks to the producers and to provide technical assistance to facilitate the producers' acceptance of the improvements.

The ultimate success of development efforts will depend on an efficient marketing infrastructure to ensure that farmers receive fair value for their sheep and, thus, are encouraged to devote time and resources to improving productivity. Marketing cooperatives could promote this efficiency and would have the additional advantage of providing an organizational basis for other types of development activities, such as distribution of genetically superior breeding stock, drugs and chemicals for flock health, and availability of credit.

CONCLUSION

We believe hair sheep are a genetic resource with considerable potential for meat production in tropical and, perhaps, temperate environments. Realization of this potential will require sustained efforts on the part of producers, scientists, and public officials to develop more efficient production and marketing systems for hair sheep. Indeed, hair sheep will be widely recognized as an important resource for food production only if these efforts are made and sustained. We trust that the information in this book will stimulate such recognition.

REFERENCES

1.   ILCA. 1979. *Trypanotolerant livestock in West and Central Africa. Vol. 1: general study.* ILCA Monograph 2. Addis Ababa, Ethiopia: International Livestock Centre for Africa.

     CARDI. 1981. *Animal production systems in the Eastern Caribbean.* Consultant Report No. 7. Morrilton, Arkansas: Winrock International.

     Mason, I. L. 1980. Prolific tropical sheep. FAO Animal Production and Health Paper 17. Rome: FAO.

2.   Scoville, O. J. and M. Sarhan. 1978. Objectives and constraints of ruminant livestock production. *World Review of Animal Production,* 14(1): 43-48.

     Fitzhugh, H. A. and A. J. DeBoer. 1981. Physical and economic constraints to intensive animal production in developing countries. A. J. Smith and R. G. Gunn (Editors). In *Intensive Animal Production in Developing Countries.* British Soc. Anim. Prod. Occ. Publ. (4):23-56.

3.   Chapters 2.4 and 2.9 in this book.

4.   Fitzhugh, H. A. 1978. Animal size and efficiency, with special reference to the breeding female. *Anim. Prod.* 27:393-401.

     Pimentel, D., W. Dritschilo, J. Krummel, and J. Kutzman. 1975. Energy and land constraints in food protein production. *Science* 190:754-761.

5.   Chapter 2.5 in this book.

6.   ILCA (1); Chapter 4.1 in this book.

7.   Bradford, G. E. 1972. Genetic control of litter size in sheep. *J. Reprod. Fert.,* Suppl. 15, pp. 23-41.

8.   Chapters 4.2 and 4.6 in this book.

9.   Chapters 2.11 and 4.2 in this book.

10. Chapters 2.1, 2.5, and 2.8 in this book.

11   Table 2.1.9 of Chapter 2.1 and Table 2.4.9 of Chapter 2.4 in this book.

12. Chapters 2.1 and 2.4 in this book.

13.   Chapters 2.2 and 3.1 in this book.
14.   Table 2.4.2 of Chapter 2.4 in this book.
15.   Table 4.1.6 of Chapter 4.1 in this book.
16.   Tables 2.10.4 and 2.10.6 of Chapter 2.10 in this book.
17.   Table 2.6.1 of Chapter 2.6, Table 2.10.1 of Chapter 2.10, Table 2.12.2 of Chapter 2.12, Table 3.1.4 of Chapter 3.1, and Table 4.6.1 of Chapter 4.6 in this book.
18.   Chapters 2.10 and 4.6 in this book.
19.   Chapter 2.10 in this book.
20.   Table 2.5.2 of Chapter 2.5 and Table 2.8.3 of Chapter 2.8 in this book.
21.   L. R. Piper, personal communication.
22.   Chapters 2.9, 4.1, 4.4, and 4.5 in this book.
23.   Chapter 2.1 in this book.
24.   Chapter 2.4 in this book.
25.   Chapter 3.3 in this book.
26.   Chapter 3.4 in this book.
27.   Table 2.4.12 of Chapter 2.4, Table 2.9.7 of Chapter 2.9, and Table 2.10.6 of Chapter 2.10 in this book.
28.   Chapters 2.1 and 2.5 in this book.
29.   Table 2.4.14 of Chapter 2.4 in this book.
30.   Table 4.1.6 of Chapter 4.1 in this book.
31.   Chapter 4.2 in this book.
32.   Chapters 2.5, 2.7, 3.1, and 3.3 in this book.
33.   Table 3.4.2 of Chapter 3.4 in this book; G. E. Bradford, personal communication.
34.   Chapter 2.7 in this book.
35.   Chapter 2.3 in this book.
36.   Turner, H. N. and S.S.Y. Young. 1969. *Quantitative genetics in sheep breeding.* 332 pp. Ithaca, N.Y.: Cornell University Press.

SECTION TWO:
MIDDLE AMERICA
AND SOUTH AMERICA

2.1.1 Virgin Island ewes, primarily white with some brown spots and black markings on muzzle, ears and around eyes, Tortola

2.1

PELIBUEY SHEEP IN MEXICO

Mario Valencia Zarazúa and Everardo Gonzalez Padilla,
Instituto Nacional de Investigaciones Pecuarias,
Secretaría de Agricultura y Recursos Hidraúlicos, Mexico

INTRODUCTION

Little is known about the origins of Pelibuey sheep in Mexico (*peli* means hair; *buey*, ox); however, similar sheep are found in West Africa. This breed has existed in Mexico for about a century or more, having been imported from Cuba.

Pelibuey sheep in Mexico are found mainly on the Gulf Coast, in the Yucatan Peninsula states, and, recently, in the Pacific Coast states. No census has been taken for this breed. However, personal communications with officials of the Secretaría de Agricultura y Recursos Hidraúlicos (SARH) in different states have yielded the following numbers: 45,000 in Veracruz; 25,000 in Yucatan; 20,000 in Tabasco; 15,000 in Tamaulipas; 15,000 in Campeche; and from 5,000 to 10,000 in each of the states of Quintana Roo, Chiapas, Guerrero, Michoacan, Puebla, and Oaxaca.

Despite the fact that Pelibuey sheep are found in almost all of the warm-humid areas of Mexico, the population is so small that its contribution to the national livestock economy is limited. Most sheep flocks constitute a secondary activity to beef producers; the breeding stock consists of 40 to 80 females and 2 to 5 males, maintained exclusively on pasture. These flocks are exploited in rudimentary form, without established programs of management, nutrition, genetic improvement, or preventive care. Marketing is inefficient and irregular (the meat usually is consumed at celebrations and parties of the owners). On the other hand, there are a few producers with flocks of 200 or more sheep as their principal occupation who use better technology and organized marketing systems. Animals from these flocks are sold mainly as breeding stock.

GENERAL CHARACTERISTICS OF PELIBUEY SHEEP

The Pelibuey breed presents a variety of solid hair colors: beige, brown, dark brown, red, white, black, and roan. The most common color combinations are brown and white; the pinto is brown with large white, well-defined spots, or white with large brown spots. Another spotted type, known as *mosqueado*, is white with small, irregular brown, red, or black spots. Less frequent types are *payaso* (clown), *golondrino* (swallow), and *panza negra* (blackbelly). The *panza negra* colors and pattern are those of the well known Barbados Blackbelly (see Chapter 1.1 for description). The *golondrino* pattern is similar to the *panza negra* pattern, but the colors are reversed so that the basic body color is dark brown or black, and the underline and other markings are lighter brown. The *payaso* has large, irregular, white or light brown spots superimposed on the *panza negra* color and pattern.

Conformation characteristics of the Pelibuey sheep include a wide and rounded forehead; the profile can be straight, semiconvex, or convex. The eye cavities are prominent with depressions behind the arches. Usually males have only horn buds, but occasionally they have horns of different length and shape; the females never have horns or horn buds. The eyes are brown or green and not very prominent, while the inner eye commisure has prominent sebaceous glands. The skin over the head is thin, tightly attached, and covered with short fine hair. The color of the mucosa inside the eyelids, mouth, and nostrils can be pink or pigmented. The ears are short and pointed.

The neck of the male is strong, rounded, short, and in the majority of cases covered with longer hair along its crest. Long hair also covers the

TABLE 2.1.1. SIZE CHARACTERISTICS OF PELIBUEY SHEEP IN MEXICO

	Study I[a]		Study II[b]		Study III[c]	
	Rams	Ewes	Rams	Ewes	Rams	Ewes
Number of animals	6	20	16	16	6	138
Adults, weight, kg	49	35	-	38.1	-	37.3
Height at withers, cm	65	66	66.3	61.2	78.6	66.8
Dorsal longitude, cm	64	56	-	-	51.2	43.4
Heart girth, cm	95	87	74.9	74.9	96.8	81.1
Metacarpus perimeter, cm	9.5	7.5	-	-	9.8	7.5

[a] Ruz (1966).
[b] Castillo, Román, and Berruecos (1972).
[c] Berruecos et al. (1975).

pharyngeal region along the lower side of the neck to the chest. In the female, this type of hair is not present, and the neck is thinner and longer. Both sexes may rarely have wattles in the pharyngeal region.

The body is slender with a straight dorsal line and slightly drooping rump. The hair is short and fine, with a less apparent inner coat of wool. The tail is thin with a white tip. The extremities are not heavily muscled, thus giving the impression of being long and angular. The hooves can be light, spotted, or totally pigmented. In the space between the hooves, there is a large sebaceous gland.

Table 2.1.1 lists some size characteristics of Pelibuey sheep.

GENERAL CHARACTERISTICS OF THE EXPERIMENT STATION AND FLOCK AT MOCOCHA, YUCATAN

Description of the Livestock Experiment Station at Mococha

Principal research on Pelibuey sheep by the Instituto Nacional de Investigaciones Pecuarias of SARH is conducted at the Mococha Station in the state of Yucatan, 24 km northeast of Merida. Unless otherwise stated, data reported here were collected on the Mococha flock. The station has 130 hectares, 70 of which are planted with introduced grasses and the rest covered with native grasses, legumes, and other native vegetation.

The soil type is calcareous with a variable layer of topsoil from 4 to 15 cm in depth. The terrain is very rocky, which makes mechanization difficult. Annual mean temperature is 21° C, ranging from 6 to 40° C. Annual precipitation is 800 mm, mostly between June and October.

The Flock and Selection Program

The original flock consisted of 250 females and 8 males and came from the experiment stations at Paso del Toro, Veracruz; Hueytamalco, Puebla; and Tizimin, Yucatan. No special criteria of selection were used in the formation of this flock, which was established in 1974. Currently, the Mococha station has 1200 females and approximately 35 males.

The selection of the rams is based primarily on weight gain to weaning (4 months of age) and weight at 6 months, 1 year, and 2 years. Initial selection is for the top 20% in weaning weight within birth type (single, twin, etc.). The final selection is made at one year of age, based on weight and conformation, good characteristics of libido, scrotal circumference, and absence of physical anomalies (such as horns, wool, split scrotum, cryptorchidism, and testicular hypoplasia).

The ewes are selected at puberty on the basis of a combination of body weight and early puberty to obtain the most precocious animals. During the productive life, ewes are selected on reproductive efficiency and maternal ability, eliminating baren ewes and those in anestrus during one season of mating. Ewes that fail to wean lambs in two successive lambing seasons, and those with mastitis and agalactia, are eliminated from the herd.

Management and Feeding

Some preliminary results on the productivity of Pelibuey sheep on improved pastures are given in Table 2.1.2, including number of animals/ha, weight gain/head, and liveweight production from pastures of Guinea (*Panicum maximum*), Ferrer (*Cynodon dactylon*), Estrella africana (*Cynodon plectostachyus*) and Buffel común (*Cenchrus ciliaris*). These data show that Pelibuey can be exploited to obtain acceptable yields/ha under intensive conditions.

The flocks at the Mococha station are pastured 4 hours in the morning (7:00 to 11:00), then corralled for water and shade. After a rest period during the hottest hours of the day (11:00 to 14:00), the animals return to pasture for another 4 hours (14:00 to 18:00). During the night ewes remain in the corral with free access to minerals and water. (Although this is the common management at the station, experimental observations indicate that higher weight gains are obtained with 24-hour grazing.)

During the dry season, February to May, when there is a natural reduction in quantity and quality of available forage in the pastures, the ewes are fed ad libitum fresh pulp of henequén (*Agave fourcroydes*) in the corral. This pulp is a by-product of the sisal fiber industries. In addition, 100 to 200 g of supplement per animal are given to lactating ewes, to weaned lambs, and to the animals grazing the lowest quality pastures. The most common supplements are sunflower, coconut, safflower, and sesame oilcake—to which molasses, sorghum, or urea are added.

Management of Flock Reproduction

Breeding Season. Pelibuey ewes exhibit estrus throughout the year, with a tendency to diminished estrous activity between February and April. Several breeding seasons are used through the year (Table 2.1.3). Each season lasts 35 days, which allows up to two matings per ewe since the estrous cycle lasts 17.5 ± 1.5 days (Valencia and Peña 1977).

Estrus Detection. During the breeding season, estrus is detected using rams whose penis has been surgically diverted. There is very little homosexual activity among ewes. Ewes are checked twice a day, morning and afternoon. There are 60 ewes assigned to each ram. At the end of the observation

TABLE 2.1.2. PRODUCTIVITY OF GRAZING PELIBUEY SHEEP

Location	Number of sheep/ha	Type of grass	Mean daily gain/head, g	Liveweight produced, kg/ha/mo	Trial length, days	Season of trial
CEP Hueytamalco[a] Puebla, INIP	14	Various	79	33.2	112	Not reported
CEP Playa Vicente[b] Veracruz, INIP	15	Panicum maximum	51	22.9	196	Feb–Aug
CEP Tizimin[c] Yucatan, INIP	17	Panicum maximum	78	39.8	196	Not reported
	17	Cynodon dactylon	67	34.2	196	Not reported
CEP Mococha[c] Yucatan, INIP	12	Cenchrus ciliaris	54	19.4	364	Year–round
CEP Tizimin[c] Yucatan, INIP	18	Cynodon plectosta-chyus	63	37.3	140	Aug–Dec

a Treviño (1977).

b Arroyo (1977).

c Torres et al. (1977).

TABLE 2.1.3.   MATING SCHEDULE FOR 1.5 LAMBINGS/EWE/YEAR

Year	Mating season (35 days)	Lambing season (35 days)	Rest period
1976	May 10 - June 15	Oct 7 - Nov 12	Oct 7 - Dec 31 (50-85 days)
1977	Jan 1 - Feb 5	June 1 - July 5	June 1 - Aug 14 (40-75 days)
1977-78	Aug 15 - Sep 20	Jan 12 - Feb 16	Jan 13 - May 9 (80-115 days)
1978	May 10 - June 15	Oct 7 - Nov 12	Oct 7 - Dec 31 (50-85 days)

MATING SCHEDULE FOR REPLACEMENT EWES

Birth season	Initial mating season	Age at first mating, range
June - July	May - June	309 - 344 days
January - February	January - February	319 - 354 days
October - November	August - September	276 - 312 days

period, detector rams are removed from the flocks to avoid overtiring and to maintain libido.

Controlled Breeding. Mating assignments are made prior to the start of the breeding season, depending on the experiments to be conducted. After estrus is detected, ewes are mated once per period (morning or afternoon) for the duration of the period. Average number of matings per estrus is three, since estrus averages 34.3 hours.

Gestation. Ewes that do not show estrus after 30 days are grouped in one herd and sent to intermediate quality pastures until the end of the second trimester of gestation. During the last trimester, the ewes are moved to better quality pastures until parturition. Anthelmintic treatment and vaccinations against septicemia and anthrax are given during the last trimester of gestation.

Parturition and Lactation. After parturition, the ewes and newborn lambs are weighed within 24 hours. The ewes must dispose of the afterbirth within 24 hours; otherwise, a treatment is given. Information is recorded for the lamb's sex, weight, color, etc.

The lambs remain in the corrals for 5 to 7 days until they are strong enough to go to the pasture with their mothers. The ewes are sent to good-quality pastures. Homogeneous herds are formed of ewes with lambs of similar age and size, thus avoiding competition for milk—a disadvantage for smaller lambs.

During lactation the lambs remain continuously with their mothers. The ewes each receive 150 g of protein supplement for up to 60 days. Ewes are observed for estrus to determine the interval between parturition and first estrous cycle.

Postweaning Management. Lambs are weaned at 3 to 4 months and placed in a separate herd. Weaned animals routinely are weighed bimonthly. At six months, males and females are separated, but kept on good-quality pastures. Females are observed for estrus to establish puberty, with first mating between 10 and 12 months of age.

Program to Obtain 1.5 Lambings/Ewe/Year

A breeding program has been established that provides for 1.5 lambings/ewe/year (Table 2.1.3). Mating and lambing seasons are each 35 days. After lambing, there are rest periods that vary according to the season of the year, with the longest rest period during the season of lower estrous activity (February–April).

REPRODUCTIVE CHARACTERISTICS OF PELIBUEY SHEEP

Age and Weight at Puberty

From a productivity point of view, it is desirable that the ewe initiate reproductive life at an early age to have the greatest number of lambs during her life. In Pelibuey sheep, there is a negative correlation (-0.16; Rodríquez 1979) between body weight and age at first estrus; precocious ewes reach puberty at heavier weight than do ewes that have first estrus at an older age. Slow-growing ewes generally reach the minimum age for puberty without having adequate body weight. These ewes must be eliminated from the herd, if precociousness is a basic criterion for selection.

Table 2.1.4 presents age and average body weight at puberty for Pelibuey ewes maintained under different management systems and at different seasons of the year. The information corresponds to ewes born in three different seasons of the year. In each experimental herd, rams were used to detect puberty.

Confined ewes, fed concentrate ad libitum, reached puberty 64 to 98 days earlier than did those on free and restricted grazing. If the type of management is not considered, average body weight is shown to be more nearly constant since 89% of the ewes weighed within a range of 21 to 23.3 kg at first-observed estrus.

For ewes born in 1978, the effect of the January–March season was obvious. From February to May 1979, no estrous activity was observed, caus-

TABLE 2.1.4. AGE AND WEIGHT AT PUBERTY OF PELIBUEY EWES[a]

Management system		Birth season		
		Oct-Nov 1976	June-July 1977	Jan-March 1978
Confinement				
No. ewes		–	42	50
Age, days	x̄	–	338.0	306.0
SD		–	27.4	103.1
Weight, kg	x̄	–	25.1	21.5
SD		–	2.4	4.7
Grazing, 24 hr				
No. ewes		114	38	67
Age, days	x̄	329.8	402.8	404.7
SD		28.5	65.8	91.7
Weight, kg	x̄	21.7	21.0	23.3
SD		2.7	2.4	3.5
Grazing, 8 hr				
No. ewes		–	–	63
Age, days	x̄	–	–	429.4
SD		–	–	99.6
Weight, kg	x̄	–	–	22.4
SD		–	–	3.2

[a] Puberty defined as first observed estrus.

ing large standard deviations in age at puberty for ewes born in the first three months of 1978.

Age and Weight at First Lambing

The age at first lambing primarily reflects herd management practices from weaning to first mating. Table 2.1.5 presents information on age and weight at first lambing for Pelibuey ewes born in different seasons and raised under different feeding and management systems. These ewes were mated at first estrus. In both confinement and free-grazing systems, most of the herd completed the replacement program shown in Table 2.1.3. Ewes under the restricted grazing system had their first lambing 1.5 to 3 months later, on the average, than did ewes under the other two systems.

In herds under the 8-hour and 24-hour grazing systems, only 23 and 55%, respectively, of the ewes were pregnant after the first mating season (when they were less than 12 months of age), indicating that a high portion of ewes under these systems had not reached puberty at that age.

These observations stress the need to establish management systems between weaning and mating that ensure ewes will reach a minimum body weight of 21 kg at 10 to 12 months of age.

TABLE 2.1.5. AGE AND WEIGHT AT FIRST LAMBING OF PELIBUEY EWES MATED AT FIRST AND SUBSEQUENT ESTRUSES

Management system	Birth season	Lambing season	Age, months			Weight, kg		
			No.	$\bar{x}$	SD	No.	$\bar{x}$	SD
Confinement	Jun-Jul/77	Oct-Dec/78	38	16.6	1.4	36	27.4	5.0
Confinement	Jan-Mar/78	Feb-Nov/79	61	16.0	3.3	61	25.8	4.2
Grazing, 24 hr	Oct-Nov/76	Feb-Apr/78	94	16.9	2.1	81	23.2	3.7
Grazing, 24 hr	Jan-Mar/78	Feb-Nov/79	56	18.5	2.8	56	26.4	4.0
Grazing, 8 hr	Jan-Mar/78	Feb-Dec/79	48	19.1	3.1	48	25.0	3.5

TABLE 2.1.6. REPRODUCTIVE EFFICIENCY OF PELIBUEY SHEEP

Mating season	Length of mating season (days)	Number of ewes exposed to ram	Ewes lambing/ ewes exposed (%)	Ewes not lambing/ewes exposed		
				Cycling (%)	Anestrus (%)	Aborted (%)
May-June/78	35	198	80.3	13.1	6.5	-
May-June/76	35	176	79.5	12.0	8.5	1.0
Jun-Jul/78[a]	35	100	80.0	18.0	1.0	1.0
Jun-Jul/78[a]	35	120	84.2	11.7	4.2	-
Jun-Aug/72	40	151	89.4	5.3	.7	4.6
Aug-Sep/75	35	172	83.7	10.4	2.9	2.9
Oct-Dec/77	35	141	86.5	13.5	-	-
Nov-Dec/71	60	152	90.8	7.9	-	1.3
TOTAL		1210				

[a] Different groups.

TABLE 2.1.7. EFFECT OF ENERGY LEVEL DURING MATING SEASON ON REPRODUCTIVE EFFICIENCY

Energy level	Number of ewes exposed	Ewes lambing/ ewes exposed %	Lambs born/ ewes exposed	Lambs born/ ewes lambing
High	112	75.9	.90	1.19
Low	101	64.3	.76	1.18
Total	213	70.4	.83	1.19

TABLE 2.1.8. EFFECT OF POSTPARTUM PERIOD ON FERTILITY IN PELIBUEY SHEEP

Number of ewes exposed	Postpartum period, days[a] $\bar{x}$	SD	Ewes lamb- ing/ewes exposed No.	%	Ewes not lambing[b] Cycling No.	%	Anestrus No.	%
124	72.2	14.5	78	62.9	18	14.5	28	22.6
186	103.2	24.3	152	81.7	23	12.4	11	5.9
89	119.8	22.4	71	79.8	16	18.0	2	2.2
120	123.2	9.4	101	84.2	13	10.8	6	5.0

[a] Calculated at the beginning of the mating season.
[b] Ewes bred that failed to conceive.

TABLE 2.1.9. INFLUENCE OF LITTER SIZE ON GESTATION LENGTH IN PELIBUEY SHEEP

Type of lambing	No.	Gestation length, days $\bar{x}$	SD
Single	936	149.5	2.3
Twin	194	149.3	2.4
Triple	10	148.9	1.6
Total	1140	149.4	2.3

Reproductive Efficiency

The reproductive efficiency of Pelibuey sheep, expressed as the percentage of ewes that lambed with respect to ewes exposed to a fertile ram, is listed by breeding seasons in Table 2.1.6.

Supplemental feed for mating ewes increased the lambing rate by 11.6%, with 0.14% more lambs born per litter. Prolificacy (litter size) in this study did not change with feed energy supplementation (Table 2.1.7).

The length of the postpartum rest period affected reproductive efficiency (Table 2.1.8). While the percentage of conception among mated ewes changed little as the rest period increased from 72 to 123 days, the incidence of anestrus diminished sharply with the increase from 72 to 103 days. Consequently, fertility (ewes lambing/ewes exposed) increased 18% when the postpartum period increased from 72 to 103 days.

Ewes with reproductive problems (such as anestrus, not pregnant after more than two matings, or aborting) are eliminated from the herd. The small percentage of ewes with these problems is replaced with young ewes.

Duration of Gestation

In Table 2.1.9, mean gestation length in days is shown according to type of lambing. Although there is little difference in gestation length among types of lambing (single, twin, or triple), there is a tendency to a shorter period as the number of lambs per litter increases. On the other hand, there is considerable variation in gestation length, with a range of 137 to 158 days (Figures 2.1.1 and 2.1.2).

Interval Between Lambing

Lambing interval is perhaps the best parameter for evaluating reproductive efficiency because it combines effects of other reproductive traits. Table 2.1.10 presents lambing intervals for Pelibuey sheep, at different lambing times, managed with controlled mating. Intervals exceed the theoretical ideal of 7.5 months because the postpartum period (Table 2.1.3) varied from 40 to 115 days, depending on the season of the year, and lactating ewes were allowed a second mating if they failed to conceive in the first mating season following parturition.

Age and Weight at Successive Matings

Table 2.1.11 shows the age and weight of Pelibuey sheep in successive mating and lambing seasons under standard management.

Number of Lambs Per Litter

The incidence of multiple lambing (twin, triple, quadruple) in Pelibuey sheep has been observed within the range of 17.8 to 39.9%, with a mean of

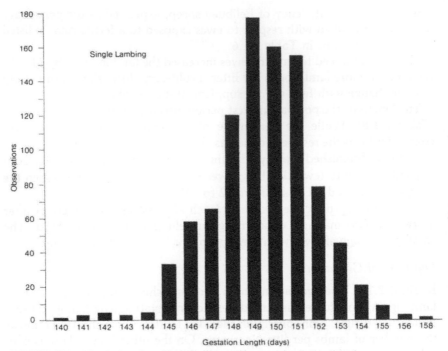

FIGURE 2.1.1 Gestation length in Pelibuey Sheep (single lambing).

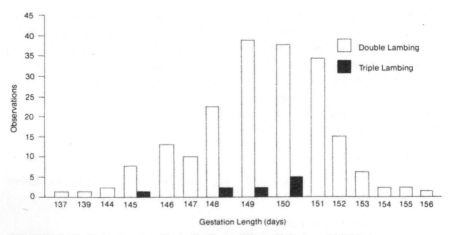

FIGURE 2.1.2 Gestation length in Pelibuey Sheep (twins and triplets).

TABLE 2.1.10. LAMBING INTERVAL FOR PELIBUEY SHEEP

Interval	No.	Length, months $\bar{x}$	SD
1st to 2nd lambing	195	9.7	2.3
2nd to 3rd lambing	92	9.8	2.9
3rd to 4th lambing	36	10.0	3.3
4th to 5th lambing	30	9.5	2.3
5th to 6th lambing	17	9.9	1.2
Total	370	9.8	2.5

TABLE 2.1.11. AGE AND WEIGHT OF PELIBUEY EWES IN SUCCESSIVE MATING AND LAMBING SEASONS

Mating season	No.	Age, months $\bar{x}$	SD	No.	Weight, kg $\bar{x}$	SD
1st	402	14.0	3.2	395	22.9	4.2
2nd	197	22.7	3.5	158	27.1	4.5
3rd	101	32.3	4.1	50	30.3	4.7
4th	35	44.4	6.8	28	33.7	3.4
5th	31	53.6	7.3	24	32.5	4.6
6th	17	61.9	5.1	7	31.4	2.7

Lambing	No.	Age, months $\bar{x}$	SD	No.	Weight, kg $\bar{x}$	SD
1st	430	18.6	3.3	428	25.7	4.6
2nd	196	27.6	3.6	190	29.5	4.6
3rd	92	37.4	4.9	87	31.8	4.0
4th	36	49.1	6.8	36	32.2	4.1
5th	30	58.1	7.0	29	34.4	5.0
6th	17	66.0	6.4	16	36.1	4.1

TABLE 2.1.12. EFFECT OF FLUSHING ON PROLIFICACY OF LACTATING PELIBUEY EWES

| Nutritional level during lactation | Start of flushing in relation to the mating season | | | | | | | | | |
| | 2 wk before | | 1 wk before | | At start | | Control | | Total | |
	No.	Born[a]	No.	Born[a]	No.	Born[a]	No.	Born[a]	No.	Born[a]
High	32	1.20	32	1.26	26	1.10	22	1.14	112	1.19
Low	30	1.26	26	1.23	29	1.06	16	1.17	101	1.18
Total	62	1.23	58	1.25	55	1.08	38	1.15	213	1.19

[a] Lambs born/ewes lambing.

TABLE 2.1.13.   PROLIFICACY IN PELIBUEY AND BLACKBELLY SHEEP

Litter size	Pelibuey No.	%	Blackbelly No.	%
One	1432	80.4	57	43.8
Two	313	17.6	58	44.6
Three	36	2.0	11	8.5
Four	0	0.0	4	3.1
Lambs born per ewe lambing	1.22		1.71	

19.6% (Valencia et al. 1975), or approximately 120 lambs born per 100 ewes lambed. These results suggest that Pelibuey sheep have low prolificacy compared to other breeds. On the other hand, 1.5 lambings/ewe/year, combined with the prolificacy index, provide for 1.4 lambs born annually per mated ewe and 1.8 lambs per ewe that has lambed.

The number of lambs per litter is influenced by the feeding regime before and after mating. Methods of supplementation (flushing) were: two weeks before and during mating; a week before and during mating; during mating; and a control with no flushing provided. It was observed that the prolificacy index increased in groups supplemented one or two weeks prior to mating (Table 2.1.12).

Prolificacy of Pelibuey and Barbados Blackbelly sheep was compared (Table 2.1.13) under similar management and feeding. Blackbelly sheep produced 0.49 more lambs per litter than did the Pelibuey at the Mococha Experiment Station.

Perinatal Mortality

The first observations on Blackbelly sheep showed a high rate (45.8%) of mortality among lambs during lactation. These data were for the first to the seventeenth week of age. Little variation was observed between sexes; 53.1% of the dead lambs were males and 46.9% were females.

However, mortality increased with the number of lambs per litter. Lamb mortality rates were 14.8, 50.5, 75.0, and 53.3% for litters of one, two, three, and four lambs respectively.

Of the 98 lambs dying during lactation, 65 (66.3%) died during the first week (Table 2.1.14). Further analysis showed that, during this first week, more than 70% of the deaths occurred in the first three days (Table 2.1.15).

For Pelibuey sheep, lamb mortality during lactation was 11.2%, with the highest incidence (5.3%, 24/453) during first-week postparturition. Stillbirths accounted for 3.1%. Pelibuey and Blackbelly flocks were managed similarly.

TABLE 2.1.14. BIRTH WEIGHT OF BLACKBELLY LAMBS DYING AT
FROM 1 TO 17 WEEKS OF AGE

Week when death occurred	Birth weight, kg					
	Male			Female		
	No.	$\bar{x}$	SD	No.	$\bar{x}$	SD
1st	37	1.80	.65	28	1.61	.58
2nd	3	1.90	.22	2	1.20	.0
3rd	2	1.86	.15	2	1.80	0
4th	2	2.00	0	2	2.20	.28
5th-17th	8	2.61	.88	12	1.93	.47
	52			46		

TABLE 2.1.15. BIRTH WEIGHT OF BLACKBELLY LAMBS DYING IN
THEIR 1ST WEEK, kg

Day when death occurred	Male			Female		
	No.	$\bar{x}$	SD	No.	$\bar{x}$	SD
Stillborn and 1st	6	1.35	.52	8	1.14	.17
2nd	12	1.82	.65	6	1.87	.89
3rd	11	1.94	.65	6	1.86	.20
4th	6	1.95	.79	2	2.15	.63
5th	–	–	–	2	1.75	.78
6th	1	2.20	0	2	1.50	.14
7th	1	1.40	0	2	1.45	.64

TABLE 2.1.16. RELATION OF SEX AND TYPE OF LAMBING TO
BIRTH WEIGHT OF PELIBUEY SHEEP, kg

Litter no.	Male			Female			Total		
	No.	$\bar{x}$	SD	No.	$\bar{x}$	SD	No.	$\bar{x}$	SD
One	717	2.74	.53	715	2.58	.49	1432	2.66	.52
Two	176	2.22	.50	168	2.07	.52	344	2.13	.48
Three	19	1.73	.63	17	1.67	.64	36	1.69	.63
Total	912	2.62	.58	900	2.47	.55	1812	2.49	.57

TABLE 2.1.17. RELATION OF SEX AND TYPE OF LAMBING TO
BIRTH WEIGHT OF BLACKBELLY SHEEP, kg

Litter no.	Male			Female			Total		
	No.	$\bar{x}$	SD	No.	$\bar{x}$	SD	No.	$\bar{x}$	SD
One	26	2.84	.90	31	2.65	.57	57	2.73	.74
Two	60	2.06	.67	56	1.95	.57	116	2.00	.63
Three	18	1.56	.46	16	1.76	.42	34	1.65	.43
Four	7	1.87	.42	7	1.64	.38	14	1.76	.16
Total	111	2.15	.80	110	2.10	.64	221	2.12	.73

High mortality rates diminish the advantages of the large litters obtained from Blackbelly sheep. Preliminary observations suggest that the main cause is malnutrition of the newborns due to lack of maternal attention.

GROWTH CHARACTERISTICS

Birth Weight Per Lambing Type and Sex

Tables 2.1.16 and 2.1.17 show birth weights by sex and type of lambing for Pelibuey and Blackbelly sheep. Males of both breeds were usually heavier than females, independent of type of lambing. Also, for both breeds, there were differences among types of lambing: single-born lambs were always heavier than those from litters of two or three. This information is consistent with that reported by Valencia et al. (1975), shown in Table 2.1.18. There is a significant relation between birth weight and weight gain to one year of age ($r = 0.48$); i.e., on the average, lambs with heavier birth weight develop faster. Lambs with birth weight of 1.1 kg or less have the highest preweaning mortality rate; thus, a factor to be considered in multiple lambing is the need to exercise special care with lighter-weight lambs.

Weaning Weight

The most critical stage of the lamb's life is postweaning, especially under grazing conditions. Between the ages of 90 and 120 days, lambs had the highest incidence of parasitic and infectious diseases, contributing to poor body development. Experiments of weaning at 75 days of age, under partial confinement, have been successful (Castillo et al. 1974); however, this method under grazing conditions has not been successful. Thus, the common practice is to wean at 3, 4, and even 5 months of age.

Tables 2.1.18, 2.1.19, and 2.1.20 show weaning weights for Pelibuey and Blackbelly sheep at 3 and 4 months of age, by sex and type of lambing.

TABLE 2.1.18. BODY WEIGHT OF PELIBUEY SHEEP AT DIFFERENT AGES[a]

Trait	Single No.	Single $\bar{x}$	Single SD	Multiple No.	Multiple $\bar{x}$	Multiple SD
Birth weight, kg	110	2.79	.48	55	2.21	.45
Weaning weight (90 days), kg	106	16.14	2.57	55	11.41	2.55
Postweaning weight (120 days), kg	92	18.07	2.90	46	12.86	2.84
Weight at 12 months, kg	91	29.25	6.81	45	23.29	6.12
	Male No.	Male $\bar{x}$	Male SD	Female No.	Female $\bar{x}$	Female SD
Birth weight, kg	88	2.71	.55	80	2.44	.52
Weaning weight (90 days), kg	86	15.32	3.54	78	13.46	3.07
Postweaning weight (120 days), kg	71	17.58	3.89	70	14.87	3.23
Weight at 12 months, kg	70	31.67	4.02	69	22.96	6.72

[a] Data collected at INIP station at Tizimin (Valencia et al. 1975).

TABLE 2.1.19. EFFECT OF SEX, TYPE OF BIRTH, AND WEANING AGE IN THE WEANING WEIGHT OF PELIBUEY SHEEP, kg

Sex	Weaning age, days	Single birth No.	Single birth $\bar{x}$	Single birth SD	Multiple birth No.	Multiple birth $\bar{x}$	Multiple birth SD
Male	90	55	11.4	1.9	19	9.5	1.9
Female	90	48	10.5	1.6	22	8.8	1.3
Male	120	54	12.7	2.4	17	10.6	1.7
Female	120	47	12.2	2.1	18	10.1	1.5

TABLE 2.1.20. WEANING WEIGHT OF BLACKBELLY SHEEP AT 120 DAYS OF AGE, kg

Litter size	Male No.	Male $\bar{x}$	Male SD	Female No.	Female $\bar{x}$	Female SD
One	19	12.4	3.0	27	10.6	2.6
Two	28	11.3	2.1	26	10.4	3.0
Three	2	15.8	.3	7	10.6	2.3
Four	3	10.6	2.2	4	8.2	2.0
Total	52	11.8	2.6	64	10.4	2.7

There were no differences in weaning weights between Pelibuey and Blackbelly lambs; however, within breed, weaning weight differences due to sex and type of lambing were substantial. Data shown in Tables 2.1.18 and 2.1.19 are from the Tizimin and Mococha flocks, respectively—which accounts for some of the large differences. Preliminary analysis indicated a significant relation between weaning weight (4 months) and weight at one year of age (r = 0.61).

REFERENCES

Arroyo, R.D. 1977. Evaluación de la capacidad de carga en pasto guinea con borrego Tabasco o Pelibuey en Playa Vicente, Ver., clima AW. *Resúmenes de la XI Reunión Anual del Instituto Nacional de Investigaciones Pecuarias*, S.A.G., p. 18.

Berruecos, V.J.M., M. Valencia Z., y H. Castillo R. 1975. Genética del borrego Pelibuey. *Téc. Pec. Méx.* 29:59–65.

Castillo, R. H., H. Román P., y J. M. Berruecos V. 1972. Crecimiento en el borrego Tabasco o Pelibuey. I. Edad y peso al destete y fertilidad de la madre. *Téc. Pec. Méx.* 21:29.

Castillo, R. H., H. Román P., y J. M. Berruecos V. 1974. Característica de crecimiento del borrego Tabasco. I. Efecto de la edad y peso al destete y su influencia sobre la fertilidad de la madre. *Téc. Pec. Méx.* 27:28–32.

Jefes de programa ganadero, 1979. Comunicación personal.

Monografía del estado de Yucatán. 1978. Gobierno del estado de Yucatán.

Rodríquez, A. R. 1979. Determinación de la pubertad y otros parámetros reproductivos en ovejas Pelibuey o Tabasco. Tesis Profesional, Escuela de Medicina Veterinaria y Zootecnia de Yucatán.

Ruz, J. G. 1966. Estudio del ovino tropical Pelibuey del sureste de México y sus cruzas con ovino Merino. Tesis Profesional, Escuela Nacional de Medicina Veterinaria y Zootecnia, UNAM, Mexico.

Torres, H. M., T. R. Garza, R. D. Arroyo, R. De León, y I. Molina. 1977. Evaluación del borrego Tabasco o Pelibuey bajo condiciones de pastoreo. *Memorias de la XIV Reunión Anual del Instituto Nacional de Investigaciones Pecuarias*, pp. 15–19.

Torres, H. M., T. R. Garza, y I. Molina. 1975. Estudio sobre capacidades decarga de borregos Tabasco o Pelibuey en zacate estrella de Africa en Tizimín, Yucatán. *Resúmenes de la XII Reunión Anual del Instituto Nacional de Investigaciones Pecuarias*, S.A.G., p. 17.

Treviño, S. M. 1977. Pruebas de aceptación de plantas forrajeras y aumento de peso con borrego Tabasco o Pelibuey en Hueytamalco, Puebla, clima AF, (c). *Resúmenes de la XI Reunión Anual del Instituto Nacional de Investigaciones Pecuarias*, S.A.G., p. 19.

Valencia Z., M., F. J. Peña T. 1977. Aspectos reproductivos del borrego Tabasco o Pelibuey. Trabajo presentado en la XV Reunión Anual del Instituto Nacional de Investigaciones Pecuarias, Jalapa, Ver. (no publicado).

Valencia Z., M., R. H. Castillo, y V.J.M. Berruecos. 1975. Reproducción y manejo del borrego Tabasco o Pelibuey. *Téc Pec. Méx.* 29:66-72.

ACKNOWLEDGMENTS

The assistance of J. M. Ponce de León C., F. J. Peña T., A. R. Rodríquez A., J. J. Méndez B., A. Velazquez M., and M. Heredia A., del Centro Experimental Pecuario de Mococha, in the collection of experimental data is gratefully acknowledged.

2.2

REPRODUCTION IN PELIGUEY SHEEP

A. Gonzalez-Reyna and J. De Alba,
Asociación Mexicana de Producción Animal, Mexico
W. C. Foote, *Utah State University, U.S.A.*

Reproductive performance was measured on Peliguey sheep managed primarily under pasture conditions near Aldama, Tamaulipas, Mexico, during 1974 and 1975. (Pelibuey and Peliguey are alternative names for hair sheep in Mexico derived from West African Dwarf stock.) The study involved 40 ewes (1 to 5 years of age and older), 2 mature rams, and lambs born during the experimental period. The rams were kept with the ewes continuously. Constants related to reproductive performance were obtained from a study of ovulation phenomena in postpartum ewes. Laparotomies were performed, and the data from laparotomized ewes were recorded separately from those of intact ewes. Ovulation studies are not detailed in this report. Tables 2.2.1 to 2.2.6 show the data obtained.

RESULTS

Ewe lambs reached puberty at an average age of 245 days; their average weight at puberty was 22.9 kg. Their mean age at first lambing was 426 days. For different samples of ewes, gestation length averaged 148.4 days.

Following parturition, an average of 42.9 days elapsed before first estrus, and an average of 46.6 days before conception. Lambing interval averaged 207.6 days, with 79% of the ewes having a lambing interval of less than 7 months.

Mean daily milk production was significantly higher ($P \leqslant 0.025$) for ewes nursing twins (630 g) than for ewes nursing singles (506 g). The mean daily weight gain of lambs from birth to 8 months was 106.6 g. The correlation coefficient between the weight at birth and weight at 8 months was 0.32.

Laparotomized ewes had longer lambing intervals, with longer anestrus after first lambings as compared to later lambings.

TABLE 2.2.1.   REPRODUCTIVE PERFORMANCE OF PELIGUEY EWES

	Number of observations	$\bar{x}$	SD	Range
Puberty				
Age, days	18	245	61.9	187–376
Weight, kg	10	22.9	2.3	17.8–25.4
Ovulation rate [a]	9	1.0	--	--
Age at first lambing	24	425.6	82.8	319–557
Lambing rates [b]				
First parturition	30	1.13	--	--
Second and later	98	1.22	--	--
Overall	128	1.20	--	--
Reproductive interval, days				
Parturition to first estrus	49	42.9	23.1	6–125
Parturition to conception	25	46.6	17.0	16–100
Gestation length	50	148.4	2.8	141–154
Lambing intervals, days				
All ewes				
First	17	213.5	38.3	181–296
Second and later	50	205.6	20.5	166–365
Overall	67	207.6	36.8	166–365
Nonlaparotomized ewes	47	198.2 [c]	19.9	166–258
First	13	198.3	18.4	181–243
Second and later	34	198.2	20.4	166–258
Laparotomized ewes	20	229.8 [d]	54.6	179–365
First	4	263	46.6	195–296
Second and later	16	221.5	54.4	179–365
Mature weight, kg				
One week postpartum	55	32.7	4.8	22.6–43

[a]   Twelve ewe lambs were laparotomized; nine had corpora lutea.
[b]   Values given as number of lambs born per ewe lambing.
[c]   198.2 vs 229.8 (P<0.01).
[d]   263 vs 221.5 (P<0.10).

TABLE 2.2.2. BIRTH WEIGHTS OF PELIGUEY LAMBS, kg[a]

Type of birth	Sex								
	Males			Females			Overall		
	No. lambs	$\bar{x}$	SD	No. lambs	$\bar{x}$	SD	No. lambs	$\bar{x}$	SD
Singles	46	2.8	.54	42	2.6	.39	88	2.7	.47
Twins	16	2.1	.32	17	2.1	.29	33	2.1	.29
Overall	62	2.6	.55	59	2.4	.38	121	2.5	.55

[a] Single lambs were heavier than twins for both sexes ($P<0.05$); single males were heavier than single females ($P<0.05$) and the combined weight of single and twin males was heavier than combined weight of single and twin females ($P<0.01$).

TABLE 2.2.3. DISTRIBUTION OF PARTURITIONS AND ESTROUS PERIODS THROUGHOUT THE YEAR

	Month												Total
	J	F	M	A	M	J	J	A	S	O	N	D	
Parturitions	7	3	15	11	6	23	12	5	8	2	5	14	111
Estrous periods[a]	10	14	5	4	9	8	3	10	5	5	18	11	98

[a] Some estrous periods were apparently not recorded.

TABLE 2.2.4. DAYS TO ESTRUS, DAYS TO CONCEPTION, GESTATION LENGTH, AND LAMBING INTERVAL FOR EWES FOR WHICH ALL MEASUREMENTS WERE AVAILABLE (22 EWES)

	Lambing interval, days			
	Days to first estrus	Days to conception[a]	Gestation length	Total
$\bar{x}$	41.5 (SE = 3.4)	47.9 (SE = 3.7)	148.2 (SE = 0.6)	195.3 (SE = 3.7)
Max	68	100	154	247
Min	9	16	142	166

[a] 77.3 percent of the ewes conceived at the first estrus after parturition (17 out of 22 ewes).

TABLE 2.2.5.   PROPORTION OF PELIGUEY EWES WITH DIFFERENT LAMBING INTERVALS, %

Type of interval	No.	Interval length, months		
		<6	6-7	>7
First-second lambings	17	0	64.7	35.3
Second-third and later lambings	50	6.0	72.0	22.0
Nonlaparotomized ewes	47	4.3	78.7	17.0
Laparotomized ewes	20	5.0	50.0	45.0

TABLE 2.2.6.   DAILY MILK YIELD OF PELIGUEY EWES DURING THE FIRST SIX WEEKS OF LACTATION, g

Days of lactation	Number of lambs nursing									
	Singles[a]					Twins				
	No.	$\bar{x}$	SD	Max	Min	No.	$\bar{x}$	SD	Max	Min
6-8	33	509	128	795	325	5	703	290	1208	470
20-22	25	573	163	1000	360	4	690	387	1265	435
42-44	33	452	145	945	242	5	510	117	700	385
Mean[a]	91	506	151	1000	242	14	630	271	1265	385

[a] Mean milk yields for ewes with singles and twins were significantly different (P<0.025).

AFRICAN SHEEP IN COLOMBIA

Rodrigo Pastrana B. and Rafael Camacho D.,
Instituto Colombiano Agropecuario, Colombia
G. E. Bradford, *University of California, Davis, U.S.A.*

ORIGIN AND DISTRIBUTION

African sheep, also known as *pelona* or *camura*, have played an important role in the animal economy of Colombia for more than 300 years. Some of the outstanding features of this breed include adaptation to hot climates, hardiness, and meat quality. However, few performance records have been kept, and there have been few selection programs or other efforts to improve these sheep. Recently, however, the Secretaria de Agricultura del Tolima and the Caja Agraria have selected some animals and organized flocks with the purpose of multiplying the sheep and selling animals to local farmers.

Because of their long history in the country and their adaptation, the African sheep often are considered native to Colombia. However, it is generally accepted that the sheep came from West Africa on ships that brought slaves to the New World. The port of entry to Colombia probably was the Peninsula de la Guajira, from which the sheep spread through the Departments of Magdalena, Atlantico, Bolivar, and Cordoba (all along the Atlantic Coast). They also are found in the Departments of Cundinamarca, Tolima, Valle, and Llanos Orientales, and in the central regions of Colombia. They are an important part of the economy of the Indian people of the Peninsula de la Guajira, where there are now an estimated 500,000 African sheep. The first wool sheep from Spain also were introduced via La Guajira, and the types probably have been crossed substantially. However, the characteristics of the African type are more evident.

In Colombia, hair sheep usually are kept in small flocks that graze the less productive areas of farms. They receive little special care and get much of their feed from browse and from scraps and by-products on the ranches. On cattle ranches, sheep (rather than cattle) often are slaughtered to provide meat for the workers. Other uses include maintenance as a source of ready cash and for feasts on special occasions. Flocks are larger in the Guajira; a large flock denotes wealth and prestige for the Indians of the area.

The African sheep have short red hair, with the color varying from deep red to light red. Both rams and ewes are polled. The ears are small and are carried horizontally. Males are maned and some animals have patches of wool on the back. Additional information is provided by Bautista (1977).

EL PALMAR FARM

The El Palmar farm was acquired by the Secretaria de Agricultura del Tolima in 1965 to develop a flock as a source of improved breeding stock for local farmers. The farm is located 348 meters above sea level and has an average temperature of 26° C and an annual rainfall of 400 mm. The sheep unit has 53 hectares planted with *Hyparrhenia rufa* and *Digitaria decumbens* (Pangola grass). A full-time shepherd is responsible for the sheep flock, under the general supervision of the farm director.

The program was started with 89 sheep: 70 ewes, 3 rams, and 16 lambs. The flock of ewes and rams graze together year-round, with no controlled breeding season. Water is available at all times from a stream running across the farm. Animals are housed each night in a shed made of bamboo where they are fed freshly cut *Bohemia nevea* and *Pennisetum purpureum*. A mineral supplement and salt are provided.

Data on performance of sheep in this flock were summarized by Montoya (1967). During the report period, the sheep were grazed on irrigated Pangola grass pasture fertilized with 150 kg/ha of urea 1 month before grazing began. Urea was sprayed every 2 months. A 13-day grazing period was followed by a 39-day rest period. At night, the animals were kept in the shed and received salt and bone meal.

Carrying capacity with this pasture scheme was estimated at 46 head/ha/year.

Mortality and Disease

Only 3 adult sheep died during the 2.5 year report period. Lamb mortality was 20% at first, due to gastrointestinal parasites (*Haemonchus, Oesophagostomum,* and *Moniezia*). After routine drenching of the lambs every 20 days until weaning, mortality was reduced to about 2%. Under present, less intensive, management, mortality is 10 to 12% in lambs and 7% in adults.

Reproduction

In 29 months, there were 287 lambs from the original 70 ewes, for an average of 1.7 lambs born per ewe per 12-month period.

TABLE 2.3.1.   REPRODUCTION AND LAMB BIRTH WEIGHTS,
                EL PALMAR FARM

Trait	Group	No.	$\bar{x}$	SD
Litter size		70	1.29	0.54
Gestation period, days	ewes with singles	50	151.7	2.58
	ewes with twins	20	152.0	1.50
Lambing interval, days	all intervals	31	213.6	36.1
	all intervals except 2 over 300 days	29	205.8	19.8
Birth weight, kg	male singles	25	2.8	0.50
	female singles	25	2.5	0.58
	male twins	13	2.2	0.63
	female twins	27	2.2	0.48

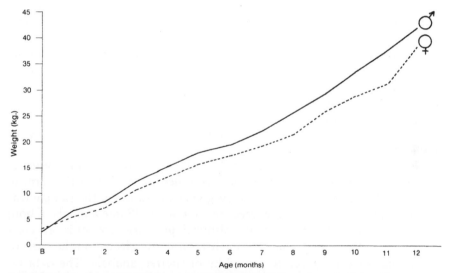

FIGURE 2.3.1 Growth of African sheep from birth to one year in Colombia.

Data on 70 lambings from 39 ewes were available for more detailed analysis and are summarized in Table 2.3.1.

The 70 lambings were recorded over a period of 24 months. Lambings occurred in 22 of the 24 months, with no noticeable seasonal pattern.

Growth Rate

Monthly weights were recorded on a sample of ram and ewe lambs from their birth through 1 year of age. Although 14 of each sex were weighed through the first 5 months, fewer were weighed for each month thereafter, with only 18 (12 females, 6 males) weighed when 1 year old. Since the weights were essentially identical up to age 5 months, all available weights were used in the growth curves in Figure 2.3.1. Under conditions of good nutrition, ewe lambs gained an average of 98 g per day and rams 109 g per day from birth to a year of age.

LA GAITANA FARM

The Caja Agraria of Colombia maintains a multiplication flock of hair sheep of two breeds at the La Gaitana farm in Huila (approximate latitude 2° N). The farm also supports cattle and horses. Elevation of the farm is approximately 1200 meters, with a mean annual temperature of 22 to 23° C, and rainfall of approximately 1500 mm. The dry months are December, January, February, and August. The sheep are on pasture year-round. They are not confined at night and remain on pasture 24 hours a day. Pastures consist mostly of seeded Guinea grass and Janeiro grass, but some are of native grasses. Brush is controlled by cutting or spraying approximately twice a year. Salt and a mineral mix are provided.

The breeds maintained at La Gaitana are African (of the type previously described) and Blackhead Persian. The flock currently numbers 100 to 110 breeding ewes of the two breeds. The flock numbered more than 1000 head prior to 1975, when numbers were reduced because of parasite and foot rot problems. The flock is used to provide rams for sale to local farmers; some ewes were transferred to other Caja Agraria farms when the numbers were reduced.

No sheep performance records are kept; however, weights and measurements on a sample of the sheep were recorded on a visit to the farm in January 1979. The sheep were in very good condition and the sample was chosen on the basis of relative freedom from wool. (Some animals of both breeds carry some patches of wool, although presence of wool is one basis for culling in the flock.) The African sheep also are selected on the basis of color, red being preferred to the white or spotted animals. The data recorded are summarized in Table 2.3.2. The samples of ewes included pregnant and open (nonpregnant), lactating and nonlactating animals.

The mean weights of mature ewes did not differ greatly between the two breeds, but the obvious differences between the breeds in height and length were confirmed by these measurements.

The high variability in weight among the African rams (SD = 9.9) is due to the weight of one 60 kg ram. This ram was not in top condition, but was simply a large animal (HG, HW, and L of 90, 75, and 80 cm, respectively). [*Editor's Note:* The presence of such a ram exemplifies a common observation in flocks of hair sheep (i.e., a high phenotypic variability), which sug-

TABLE 2.3.2.　　WEIGHTS AND BODY MEASUREMENTS OF AFRICAN AND BLACKHEAD PERSIAN SHEEP, LA GAITANA, COLOMBIA

Breed	Sex	Age, yr	No.	Weight kg $\bar{x}$	SD	Heart girth cm $\bar{x}$	SD	Height at withers cm $\bar{x}$	SD	Length cm[a] $\bar{x}$	SD
African	F	2-7	15	36.7	2.9	78.0	3.1	62.9	2.2	65.1	2.0
		1.5	3	33.7	2.5	76.0	1.0	63.7	4.0	63.0	1.7
	M	3	3	48.7	9.9	84.0	5.3	68.7	5.5	73.0	6.1
Persian	F	2-8	11	35.6	2.7	78.8	3.4	60.1	1.7	61.1	3.2
		1.5	1	38	-	78	-	63	-	59	-
	M	6	1	50	-	87	-	63	-	66	-

[a]　From point of shoulder to pin bone.

gests the possibility of producing fairly rapid change through selection. In another flock in northern Colombia, two rams had heart girths of 92 and 93 cm and on this basis had estimated weights of 60 to 65 kg.]

There were no records on reproduction of the La Gaitana flock. The twinning rate of the African breed was reported as not more than 10%; the number of sets of twins observed in a group of recently lambed ewes was consistent with this percentage. Apparently, twins are extremely rare in the Blackhead Persian flock, which is in agreement with other data on this breed (Bodisco et al. 1973).

Inventory records, in which the sheep were classified by sex and age categories, were available for this flock for a period of several years. Due to sales of animals (including mature breeding ewes), reproductive rates could not be estimated accurately. However, these data suggest that: (1) the weaning rate for African sheep appeared to be appreciably less than one lamb per ewe per year and (2) the weaning rate of the Blackhead Persian flock was consistently 50% or less than that of the Africans. (These estimates support an observation made when examining the sheep: the flock has a number of nonreproducing ewes. This observation holds particularly among the Persian group, which included ewes up to 4 years of age that apparently had never lambed. A program of recording on production performance and culling would no doubt improve net reproductive rates in these sheep.)

The weights recorded on ewes 2 years old and older (36.7 kg for the Africans and 35.6 kg for the Blackhead Persian) are probably close to maximum mature weight for these strains of sheep. This conclusion is based on (1) their continuous access to good quality pasture and (2) the apparent good condition and health of all the sheep. A number of the mature Persian ewes were very fat (Photo 1.1.37), and the mean weight of this sample may be somewhat higher than optimum for the strain.

Figure 2.3.1 and Table 2.3.2 indicate that, in general, the Colombian strain of African sheep appears to be characterized by very good adaptability to a tropical environment, low prolificacy relative to several other strains of this type, and possibly somewhat heavier weights than other strains. The Blackhead Persian sheep appear to have reproductive problems, possibly as a result of inbreeding, and do not have any obvious advantage over the African type.

REFERENCES

Bautista, Riberto. 1977. *Manual de ovinos—oveja Africana.* Temas de Orientación Agropecuaria No. 125. 64 pp. Bogotá, Colombia.

Bodisco, V., C. M. Duque, y A. Valle S. 1973. Comportamiento productivo de ovinos topicales en el periodo 1968–1972. *Agronomía Tropical* 23(6):517–540.

Montoya, Rafael. 1967. Estudio sobre la oveja Africana. Tesis de Grado. 89 pp. D.M.V.Z. Facultad de Veterinaria. Universidad Nacional de Colombia. Bogotá, Colombia.

COMMERCIAL HAIR SHEEP PRODUCTION IN A SEMIARID REGION OF VENEZUELA

Carlos González Stagnaro, *Instituto de Investigaciones Agronómicas, Facultad de Agronomía, Universidad del Zulia, Venezuela*

INTRODUCTION

The semiarid tropics of western Venezuela are characterized by high temperatures and limited, irregular rainfall. Some of the extensive areas of grazing land are better suited to the grazing of sheep than of cattle. Hair sheep, known in Venezuela as "West African" or "Red African," appear to offer considerable potential for increased production.

Sheep distribution has spread in this area, due to their apparent adaptability to a hot climate (either humid or arid), resistance to diseases, hardiness, and ability to withstand great seasonal variation in forage quality. Other advantages of these sheep are their precocity, fertility, maternal ability, and their ability to breed throughout the year. A profitable system for raising these sheep and managing their reproduction can be developed on the basis of systematic study of the production characteristics of the breed under tropical conditions.

This study was designed to provide information on the reproductive physiology, to establish productive capabilities, and to outline methods of management for Red African sheep.

ORIGIN AND CHARACTERISTICS OF THE AFRICAN BREED OF HAIR SHEEP

Until a few years ago, the majority of the small numbers of sheep in Venezuela were of the Criollo type. These sheep do not constitute a distinct breed; however, their external characteristics of white fleece with spots (especially black spots), presence of horns, and long coarse wool indicate that they descended from Spanish Churra and Lacha breeds brought originally by the conquistadors. Criollo sheep have a fleece of uneven wool mixed with coarse hair, a relatively low growth rate, poor meat conformation, and low economic value.

The first attempts in Venezuela to develop sheep production were conducted with insufficient technical information and usually involved introducing purebreds of the wool breeds. Thus, the productive potential of these sheep was overshadowed by their poor performance, the lack of a pastoral vocation in Venezuela, the absence of controls that could identify causes of failure, and the near-total loss of the importations.

Recently, Criollos have been crossed with animals of the Red African type. Although the Red African had not been selected previously, they appear to have improved producing ability and the crosses are favored in peasant flocks. Progressive growers with capital have participated in more intensive exploitation to obtain maximum productivity and economic production.

These sheep of West African origin are found with similar characteristics throughout the Caribbean and equatorial America, but have different names in the different countries. In Venezuela, their numbers have grown spectacularly. In the past decade, they have come principally, but not exclusively, from Colombia (Tolima, Monteria, Valle Dupar, and Guajira), spreading from west to east. Conservative estimates indicate an increase of 200,000 head since 1969, with a 65% increase from 1975 to 1979. The state of Zulia is the leading producer with 112,000 head, or 55% of the national total.

African hair sheep are characterized as dual purpose: they provide both meat and skins. However, in selection for their use in the tropics, consideration must be given to critical traits such as prolificacy, milk yield, maternal ability, and survival rates of the young. These animals are generally healthy, active animals of hardy constitution.

Coloring can be classified into three general groups: (1) various shades of red, (2) tan or yellow, and (3) white. In classes 1 and 2, the head and belly are usually lighter colored than the rest of the body. In crossbred animals, spots are common —including, particularly, a white tip on the tail. The skin is covered with short hair similar to that of goats. Patches of wool found on the shoulders and back indicate crossing with wool breeds or with Criollos. Adult males usually have a mane.

PERFORMANCE OF THE SHEEP

From 1974 to 1979, West African hair sheep were studied in both purebred and crossbred commercial breeding flocks located in the semiarid region near the city of Maracaibo (10° 31′ N, 71° 50′ W). Average minimum and maximum temperatures for the year are 23.4° and 33.9° C, with a mean annual temperature of 27.2° C. Diurnal variation ranges from 3° to 13° C.

Relative humidity is usually between 74 and 79% and average radiation between 442 and 530 cal/cm²/day. Rainfall varies from 0 to 10 mm in the months of January and February and 145 to 148 mm in two rainy seasons, about May and October. Difference in day length between the longest and shortest days is 72 minutes.

The system of management is semiextensive, with irregular and variable pasture, including a few improved and managed pastures; most pastures, however, are native grasses including Guinea (*Panicum maximum,* Jacq), Pangola (*Digitaria decumbens*), and buffel (*Cenchrus ciliaris*) on poor, acid soils of coarse, permeable sand.

Grazing usually is allowed from 6 to 11 A.M. and from 2 to 5 P.M. for a total of 6 to 8 hours. Feeding is supplemented with minerals and salt. Occasionally, the sheep are fed concentrates during the dry season (January–April) and, rarely, in late gestation and early lactation.

Mature Weight

Weights and some body measurements were recorded on 117 rams and ewes of a well-managed commercial flock of selected Red African sheep. Age was determined by examination of the teeth.

Reproduction

Reproductive performance was evaluated for several hundred ewes in two flocks (one pure and the other crossbred), between 1976 and 1978. Characteristics studied included: age at first estrus, estrous activity, fertility, gestation period and postpartum interval in relation to season, type of feeding, and management. The effects of type of birth (single, twin, etc.) on age at puberty and on postpartum interval also were examined. A plan of programmed reproduction was evaluated involving estrus synchronization and either natural mating or artificial insemination.

The frequencies of natural cycles and of parturitions were studied in 4 flocks of African or African crossbred sheep. A total of 1183 observations were recorded on 306 ewes. The average weight of 132 ewes 8- to 14-months old was 21.6 kg; for 182 ewes 2- to 6-years old, the average was 28.6 kg. These ewes were with rams (3 to 5 rams per 100 ewes) year-round. Incidence of estrus, lambing interval, litter size, and birth weight were recorded and related to hours of daylight, temperature, humidity, and rainfall. The seasonality of the breeding pattern was evaluated from the percentage of animals coming in heat each month and from lambing intervals at different seasons of the year.

Lamb Mortality and Maternal Behavior

During periods of 6 to 20 months in 1974–1975, data on parturition were recorded in 5 flocks of predominantly African breeding located in different parts of the state of Zulia. Records were made of lamb mortality by period: first week; weeks 2 to 4, 4 to 8; months 3 to 4, 5 to 6 and 7+. The flocks were considered typical of the area flocks under semi-intensive management. Records were kept on birth weight, sex, type of birth, breed or cross, age, and, in some cases, type of illness, date, and cause of mortality.

Birth Weight, Growth Rate, and Dressing Percentage

Two flocks of West African sheep crossed with Criollo and Blackhead Persian were evaluated for birth weight, growth rate, and dressing percentage. Effects of type of birth and sex also were evaluated. The lambs and their dams usually were confined for the first 3 weeks. The ewes were fed guinea grass or forage sorghum and 300 g/d of a 14% protein concentrate, and then pastured on guinea grass.

RESULTS AND DISCUSSION

Weights and some body measurements recorded in a flock of West African sheep selected for uniform color and freedom from wool are presented in Table 2.4.1. Mature ewes reached a weight of 40 kg and mature rams weighed 57 kg.

Tables 2.4.2 and 2.4.3 show the estrous cycle traits recorded in 3 flocks. Mean cycle length was 17 days for ewes of all ages (Table 2.4.2) indicating that these sheep are very similar to temperate zone breeds in this respect. The cycle was slightly shorter (16.8 days) in young ewes than in mature ewes (17.2 days). The cycle also was shorter in the dry season (16.8 days). Duration of estrus (Table 2.4.3) was shorter (24.4 hours) in young ewes than in mature ewes (26.7 hours). The length of the estrous period of ewes that became pregnant (26.1 hours) did not differ from that of ewes returned to estrus (26.8 hours). This period for naturally mated ewes did not differ from that of ewes inseminated artificially.

Age and weight at puberty (first estrus) were measured in different groups of ewes. Ewes raised with better levels of management (Table 2.4.4) were younger at puberty as were ewes raised as singles compared to those raised as twins (Table 2.4.5). Other data from this project (not shown in the tables) indicate that lambs born from October to January reach puberty earlier than do those born during other months of the year.

As a general management practice, it is recommended that ewe lambs should not be bred at first estrus but should be older and closer to their mature size before being bred. The utility of this principle is demonstrated by Table 2.4.6, which shows that ewes that were heavier at first service had higher lambing rates and lower lamb mortality.

TABLE 2.4.1.　　BODY WEIGHT AND MEASUREMENTS OF WEST AFRICAN SHEEP ACCORDING TO SEX AND AGE (FLOCK D)

Sex	Age class	No.	Weight kg		Body measurements cm							
					Ht. at withers		Chest depth		Heart girth		Length	
			$\bar{x}$	SD	$\bar{x}$	SD	$\bar{x}$	SD	$\bar{x}$	SD	$\bar{x}$	SD
Ewes	Milk-teeth	14	31.0	4.1	61.8	3.4	27.9	1.8	74.1	4.4	60.0	5.0
	2-teeth	34	37.1	5.4	64.6	3.5	28.6	3.3	78.3	4.9	62.6	6.3
	4-teeth	21	38.8	5.4	63.8	3.1	29.8	2.6	80.4	6.2	62.6	3.3
	6-teeth	16	39.2	6.9	65.6	3.5	29.8	3.3	81.1	4.9	63.9	5.6
	Aged	21	40.7	6.2	65.2	3.7	32.3	2.8	82.8	6.0	65.0	5.3
Rams	Mature	11	57.0	11.5	69.9	7.0	35.1	3.6	92.4	8.4	68.9	6.0

TABLE 2.4.2. ESTROUS CYCLE LENGTH IN YOUNG AND MATURE
WEST AFRICAN EWES (FLOCKS A, D, E)

Cycle length, days	Ewe lambs		Mature ewes		All ewes	
	No.	%	No.	%	No.	%
13	6	2.4	3	0.9	9	13
14	8	3.2	14	3.9	22	3.2
15	12	4.8	39	10.8	51	8.8
16	89	35.7	101	28.1	190	27.5
17	56	22.5	87	24.2	143	20.7
18	39	15.7	53	14.7	92	13.3
19	25	10.0	21	5.8	46	6.7
20	8	3.2	13	3.6	21	3.0
21	4	1.6	9	2.5	13	1.9
22	2	0.8	20	5.6	22	3.6
No.	249		360		609 (578)[a]	
Mean	16.79		17.24		17.06 (16.90)	
SD	2.19		2.54		3.09 (2.47)	
Range	4-25		10-32		4-32 (14-21)	

[a] Values in () for ewes with cycles of 14-21 days only.

TABLE 2.4.3. DURATION OF ESTRUS IN YOUNG AND MATURE
WEST AFRICAN EWES (FLOCKS A, D, E)

Length of estrus, hr	Ewe lambs		Mature ewes	
	No.	%	No.	%
< 10	5	2.3	10	2.4
10-14	11	5.1	24	5.8
14-18	16	7.4	31	7.5
18-22	48	22.2	56	13.6
22-26	52	24.1	93	22.5

(cont.)

TABLE 2.4.3.   (cont.)

Length of estrus, hr	Ewe lambs No.	Ewe lambs %	Mature ewes No.	Mature ewes %
26-30	39	18.1	70	16.9
30-34	27	12.5	43	10.4
34-38	12	5.6	41	9.9
38-42	6	2.8	21	5.1
42-46	0	-	11	2.7
>46	0	-	13	3.1

	All	Ewe lambs	Mature ewes
	No.	216	413
	Mean, hr	24.4[a]	26.7[b]
	SD	6.9	9.2

[a,b] Significantly different ($P < 0.01$)

TABLE 2.4.4.   AGE AND WEIGHT AT PUBERTY OF WEST AFRICAN EWE LAMBS

Flock	No. ewes	Age, days $\bar{x}$	Age, days SD	Age, days Range	Weight, kg $\bar{x}$	Weight, kg SD	Weight, kg Range	% of mature wt $\bar{x}$	% of mature wt Range
A, D, E	196	286	41.3	190-420	20.9	2.4	13.6-23.6	61	40-69

Flock	Treatment group	No. ewes	Age, days $\bar{x}$	Age, days SD	Weight kg	% conceived 1st service	Litter size born	% lamb mortality
D	Pasture only	20	305	47	20.6	80	1.12	17
	Pasture + supplements + parasite control	19	269	31	22.1	89.5	1.29	21

TABLE 2.4.5.   AGE AT PUBERTY AND FIRST LAMBING IN
                    WEST AFRICAN EWE LAMBS IN RELATION TO
                    TYPE OF BIRTH (FLOCKS A, D)

Type of birth	Age at puberty, days			Age at first lambing, days		
	No.	$\bar{x}$	SD	No.	$\bar{x}$	SD
Single	98	262	32.6	93	418	34.3
Multiple	72	312	53.1	69	463	52.8

Reproductive performance was much higher for matings in the rainy season than for those in the dry season (Table 2.4.7). Conception rate and litter size were both higher, and the interval from parturition to the next estrus and conception was lower, due to the better nutrition available during the rainy season. The effect of lactational status (Table 2.4.7) probably is also a reflection of the condition of the ewe—fertility and litter size *decreased* as the number of lambs being suckled at the time of mating *increased*.

Fertility and litter size tended to increase as the interval from parturition to conception increased from less than 30 days to about 60 days; however, there was no further increase in interval to first service. The interval from parturition to conception was significantly shorter for the ewes with shorter intervals to first service (Table 2.4.8).

Table 2.4.9 shows data on gestation periods. First-lambing ewes had significantly shorter gestation periods than did mature ewes.

Litter size was increased from 1.14 to 1.72 by programmed reproduction, which involved estrus synchronization and the use of PMSG (Table 2.4.10). The shorter gestation period of ewes under programmed reproduction is probably an effect of the larger litter size.

Seasonality of Breeding

Breeding season in this environment is influenced both by natural factors, such as variation in nutrition as a result of rainfall, and by management imposed (González et al. 1979a).

The interval between parturitions under a system of continuous exposure to males averages about 8 months (Table 2.4.11), with high fertility and parturition rates from matings 2- to 3-months postpartum (Table 2.4.8). Prolificacy is low, but flocks vary widely. The low prolificacy may be caused by the low level of nutrition and possibly by inbreeding.

Table 2.4.11 shows that the seasonal pattern of breeding involves a prin-

TABLE 2.4.6.   REPRODUCTIVE PERFORMANCE OF WEST AFRICAN EWE LAMBS ACCORDING TO AGE
AND WEIGHT AT FIRST SERVICE (FLOCKS A, D)

Weight at 1st service kg	No. ewes	Age, days $\bar{x}$	Age, days SD	% conc. at 1st service	Lambs born per ewe lambing	Lambs born per ewe bred	Lamb mortality %	Days from parturition to 1st service
< 18	27	245	33.0	63.0	1.06	.67^a	33.3	48.3
18-22	68	251	27.9	70.6	1.06	.75^a	21.6	45.9
22-26	63	298	39.4	92.1	1.10	1.02^b	10.9	63.1
> 26	28	336	21.8	82.1	1.13	.93^b	7.7	68.6
All	186			78.5	1.09	.85	16.3	56.6

a,b Means followed by different letters are significantly different (P<0.05).

TABLE 2.4.7.   EFFECT OF SEASON AND LACTATIONAL STATUS ON REPRODUCTION IN WEST AFRICAN EWES (FLOCKS A, D)

	% fertility		Litter size		Interval, days				Services per conception	
					Part. 1st serv.		Part. conc.			
Variable	No.	$\bar{x}$	No.	$\bar{x}$	No.	$\bar{x}$	No.	$\bar{x}$	No.	$\bar{x}$
Season when mating occurred:										
Dry	169	80.5	157	1.04	135	85.2	115	107.9	115	1.23
Rainy	316	94.3	308	1.17	298	61.8	281	78.3	281	1.12
Lactational status:										
Dry	89	94.0	101	1.23	75	45.8	67	59.1	67	1.14
Nursing 1 lamb	119	91.6	135	1.15	111	58.6	103	74.3	103	1.17
Nursing 2-3 lambs	147	83.0	153	1.08	120	85.5	108	103.0	103	1.27

TABLE 2.4.8.   REPRODUCTIVE EFFICIENCY IN RELATION TO INTERVAL BETWEEN PARTURITION AND FIRST SERVICE (NATURAL MATING) (FLOCKS A, D)

Interval from part. to 1st service, days	% fertility	Litter size	Interval from part. to conception, days
< 30	75.0	1.05	36.6
30-45	82.5	1.09	56.4
45-60	88.2	1.21	73.0
60-90	93.1	1.15	89.1
90-120	91.6	1.12	111.3
> 120	82.9	1.14	132.9
All	88.6	1.14	86.9

TABLE 2.4.9.   GESTATION PERIOD FOR FIRST AND SUBSEQUENT PARITIES OF MATURE EWES IN RELATION TO TYPE OF BIRTH AND SEX OF SINGLE LAMBS (FLOCKS A, D, E)

Parity	Sex of lamb	Singles No.	Singles $\bar{x}$	Twins No.	Twins $\bar{x}$	All ewes No.	All ewes $\bar{x}$ (range)	SD
First	M	80	149.8					
	F	71	149.4					
	Both	151	149.6[a]	64	148.7[a]	215	149.2 (143-153)	2.5
Second +	M	114	151.5					
	F	107	151.1					
	Both	221	151.3[b]	185	150.6[b]	406	150.9 (141-159)	2.8

[a,b] Within a column, means followed by different letters are significantly different (P<0.05).

TABLE 2.4.10.   DISTRIBUTION OF LITTER SIZES IN EWES IN TRADITIONAL AND SYNCHRONIZED REPRODUCTION PLANS (FLOCKS A, D, E)

Program	Type of insemination	Type of birth								Total no. ewes	Mean litter size	Gestation period	
		Single		Twin		Triplet		Quad				$\bar{x}$	SD
		No.	%	No.	%	No.	%	No.	%				
Traditional	Natural	248	86.4[a]	38	13.2[a]	1	0.3[a]	-	-	287	1.14	150.8	2.1
Synchronized	Natural	92	35.9[b]	136	53.1[b]	24	9.4[b]	4	1.6	256	1.77	149.7	3.0
Synchronized	A.I.	32	41.0[b]	40	51.3[b]	5	6.4[b]	1	1.3	78	1.68	150.2	2.6

a,b Within a column, percentages followed by different letters are significantly different ($P < 0.05$).

TABLE 2.4.11.   SEASONAL VARIATION IN INCIDENCE OF ESTRUS AND LAMBING FOR WEST AFRICAN AND WEST AFRICAN CROSSBRED EWES (FLOCKS A, C, D, E)

| | Ewes in estrus | | Mean litter size | Lambing interval, days | |
	No.	%		$\bar{x}$	SD
December–March	89	7.5[a]	1.19	228.8	36.5
April–June	622	52.6[c]	1.12	244.6	18.2
July–August	76	6.4[a]	1.23	276.6	31.9
September–November	396	33.5[b]	1.16	216.0	21.9
All	1183	100.0	1.14	235.9	28.8

[a,b,c] Within a column, percentages followed by different letters are significantly different ($P<0.05$).

cipal season between April and June and a secondary season between September and November; these two periods coincide with the major and secondary periods of rainfall. This relationsip between rainfall and incidence of estrus is shown clearly in Figure 2.4.1. The duration of postpartum anestrus is also affected by the number of lambs suckled (Table 2.4.7).

Lamb Mortality

Table 2.4.12 shows the incidence of lamb mortality and factors affecting it in flocks of West African sheep and West African crosses. Mortality on 5 ranches averaged 25%, but varied from 13 to 35% between ranches. Of the 180 lambs that died in this study, the percentages of total loss were: first week (34%), weeks 1 to 4 (19%); and months 1 to 2 (4%), 2 to 4 (14%), 4 to 6 (16%), and 6 to 8 (13%). Mortality was higher ($P<0.05$) for males than for females, and for twins than for singles.

Birth weight and mortality showed a strong negative relationship (Table 2.4.13). Average birth weight (1.9 kg) of lambs that died was significantly less than that of all lambs (2.3 kg). Major causes of death up to age 8 months included (as percentage of total mortality): intestinal parasites, 19.2%; coccidiosis, 12.6%; contagious ecthyma, 8.6%; and lung worms and starvation, 7.1% each. Lamb mortality decreased with age of the dam: 36.8% for lambs from first-lambing ewes, 28.0% for lambs from ewes 2 to 3 years old, and 17.6% for lambs from mature ewes, based on a study of 640 lambs and their dams.

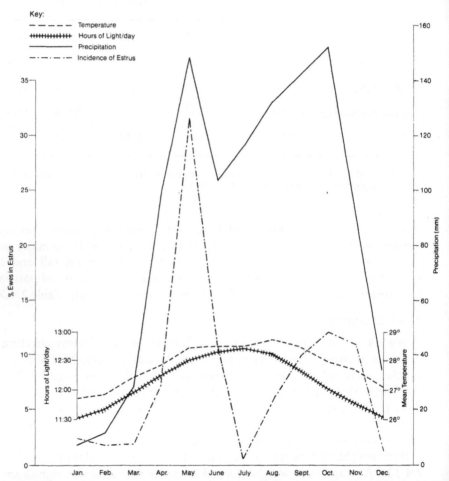

FIGURE 2.4.1 Relationship between the incidence of estrus and precipitation, hours of daylight, and temperatures in crossbred ewes in a tropical environment (10°/N, 72°/W), 1974.

TABLE 2.4.12. LAMB MORTALITY FROM BIRTH TO 8 MONTHS ON 5
RANCHES IN THE STATE OF ZULIA (1974-75), IN
RELATION TO BIRTH WEIGHT, SEX, AND TYPE
OF BIRTH

		Lambs recorded	Total mortality %	Birth weight, kg	
				No.	$\bar{x}$
Ranch:	A	318	28.0	286	2.18
	B	82	35.3	71	2.29
	C	99	24.2	132	2.31
	D	142	18.3	116	2.31
	E	90	13.3	80	2.62
	All	731	24.6	685	2.29
Sex:	Male	417	27.1		
	Female	314	21.3*		
Type of birth	Single	428	21.5		
	Multiple	303	29.0**		

* M and F significantly different (P<0.05).

** Singles and multiples significantly different (P<0.01).

There were large differences in lamb mortality among breeds and between purebreds and crossbreds in a flock that maintained several different
groups (Table 2.4.14). Criollo and crossbred lambs had low mortality, and
imported wool breeds had extremely high mortality.

Lamb mortality was reduced significantly on 3 ranches by improving
management and nutrition during pregnancy. During 1974–1975, a total of
550 lambs suffered 23.1% mortality; with improved management, 516
lambs on the same 3 ranches in 1976–1977 averaged only 11.2% mortality.

The effects of maternal age, nutrition, and behavior on lamb mortality
have been studied and published elsewhere (González, 1976, 1978).

Birth Weight and Postnatal Growth

Birth weights were recorded over a two-year period for lambs from flocks
with controlled matings maintained on pasture (Table 2.4.15). The flock's

TABLE 2.4.13. MORTALITY DURING THE FIRST WEEK IN RELATION TO BIRTH WEIGHT (FLOCKS A, B, C, D, E)

Birth weight, kg	No. lambs recorded	% mortality Birth – 24 hrs	2-7 days	Total No.	%
< 1.0	25	40.0	28.0	17	68.0
1.0-1.5	54	13.0	16.7	16	29.6
1.5-2.0	170	1.8	2.9	8	4.7
2.0-2.5	229	1.7	2.2	9	3.9
2.5-3.0	181	0.6	2.8	6	3.3
> 3.0	43	-	2.3	1	2.3
Not recorded	29	-	-	4	13.7
Total	731	4.0	4.4	61	8.3

TABLE 2.4.14. LAMB MORTALITY FOR DIFFERENT BREEDS AND CROSSES IN THE SAME COMMERCIAL FLOCK (1974-1977)(FLOCK A)

Breed or cross	No. lambs	No. died	% mortality
Wool breeds (Suffolk, Rambouillet, Corriedale)	37	24	64.9[a]
West African	64	15	23.4[b]
West African crosses	139	23	16.5[c]
Criollo	114	16	14.0[c]
Blackhead Persian (PCN)	14	5	35.7[b]
PCN crosses	86	9	10.5[c]
Total	454	92	20.3

[a,b,c] Percentages followed by different letters are significantly different (P<0.05).

TABLE 2.4.15. BIRTH WEIGHT OF WEST AFRICAN LAMBS BY TYPE OF BIRTH AND SEX (FLOCK D)

Sex	Single			Twin			Triplets			Quad.			Mean		
	No.	$\bar{x}$ (kg)	SD	No.	$\bar{x}$ (kg)	SD	No.	$\bar{x}$ (kg)	SD	No.	$\bar{x}$ (kg)	SD	No.	$\bar{x}$ (kg)	SD
Male	171	2.64	.61	122	2.17	.52	20	1.64	.54	6	1.18	.46	319	2.36	.58
Female	195	2.60	.58	126	2.08	.56	28	1.73	.50	10	1.48	.41	359	2.32	.55
Total	366	2.62	.54	248	2.12	.58	48	1.68	.50	16	1.38	.44	678	2.34	.56

TABLE 2.4.16. GROWTH OF WEST AFRICAN LAMBS ON PASTURE IN A COMMERCIAL FLOCK, BY SEX AND TYPE OF BIRTH (FLOCK D)

Age	Sex	Weight, kg										
		Singles			Twins			Triplets			Overall		
		No.	x̄	SD	No.	x̄	SD	No.	x̄	SD	No.	x̄	SD
10 days	M	45	4.6	.8	17	3.4	.9	3	2.2	1.0	65	4.2	.9
	F	80	4.0	.8	43	3.2	1.0	8	2.4	1.1	131	3.6	.9
30 days	M	39	7.9	1.4	11	5.4	1.6	3	3.8	1.7	53	7.1	1.5
	F	76	6.4	1.3	34	5.1	1.5	6	3.4	1.6	116	5.9	1.4
90 days	M	22	15.9	2.3	8	11.0	2.5	3	9.0	2.7	33	14.1	2.5
	F	57	11.4	2.4	26	8.1	2.6	4	4.4	2.7	87	10.1	2.6
120 days	M	28	16.9	3.1	7	11.6	3.3	3	9.0	3.8	38	15.3	3.5
	F	60	13.9	3.3	20	10.9	3.4	–	–	–	80	13.2	3.4

TABLE 2.4.17.   GAIN AND DRESSING PERCENTAGE OF 6-8 MONTH OLD WEST AFRICAN WETHER LAMBS FINISHED UNDER DIFFERENT FEEDING SYSTEMS (FLOCK D)

Feeding system	No. lambs	$\bar{x}$ initial weight kg	$\bar{x}$ final weight kg	$\bar{x}$ daily gain g	$\bar{x}$ dressing %
Pasture (Guinea grass)	9	21.1	27.2	77[b]	40.8[b]
Semi-confinement (Guinea grass, concentrate)	9	19.7	31.8	172[a]	43.2[b]
Confinement with concentrate	9	18.9	28.8	136[c]	47.2[a]

[a,b,c] Means within a column followed by different letters are significantly different $(P < 0.05)$.

overall average birth weight per lamb was 2.34 kg. Female lamb weights were 98% of male weights. Average weight for twins was 81% of the weight for single lambs; for triplets, 64%; and for quadruplets, 53%.

Females gained less rapidly than males, weighing 82% and 72% as much as males at 30 and 90 days (weaning age), respectively (Table 2.4.16). The differences in rate of gain between males and females in the 90- to 120-day period reflect different postweaning treatment for the two groups (González et al. 1979b). Type of birth had an even greater effect on gain, with twins weighing an average of only 73% as much as singles at 120 days. The weights of the small number of lambs raised as triplets indicate that these ewes did not have sufficient milk to support adequate growth.

Table 2.4.17 shows the growth rate and dressing percentage of a small group of high-grade West African wether lambs (castrated males) subjected to 3 different finishing methods of feeding. Gains of up to 177 g per day for 70 days were achieved by lambs that averaged 212 days of age and 20 kg at the start of the test. Dressing percentages were relatively low: 41% to 47%. Hernández (1977) reported dressing percentages in the range of 45 to 49% for West African-type sheep of different sex and age classes.

SUMMARY

Several studies are reported dealing with the physiology and reproductive performance of West African sheep in a semiarid region of Venezuela (Zulia State).

The results indicate some seasonality in breeding activity of the ewes in this environment, probably due to nutritional factors and a short lactational anestrus. Fertility, prolificacy, and lamb survival were evaluated under systems of traditional and controlled reproduction. Growth rate was measured under different combinations of pasturing and feed supplementation. West African sheep were shown to have good fertility and to be capable of lambing every 8 months under semi-extensive conditions. Twinning rate is only 10 to 20% under these conditions, but can be increased by hormone treatments. Lamb mortality is about 20% among single lambs under average management, and much higher among multiples; it can be decreased significantly by improvements in feeding and management. Lamb mortality is much lower for West African sheep than for imported wool breeds in this environment, and also is lower for the West African sheep than for the Blackhead Persian breed.

The West African sheep seem well-adapted to this environment and show good potential for improved systems for production of animal protein in a tropical environment.

REFERENCES

González, C. 1976. Ovinos tropicales: La oveja Roja Africana. Copia multigraf. 19 pp. Fac. Agronomía, Univ. Zulia, Maracaibo.

González, C. 1978. Comportamiento maternal y supervivencia de las crías en ovejas Roja Africanas. Symp. Etología Animal. XXVIII Conv. Anual Asoc. Venez. Avance de la Ciencia, Maracay.

González, C., J. Goicochea, and F. Perozo. 1979a. Comportamiento y eficiencia reproductiva en ovejas de pelo West African en el medio tropical. XXIX Conv. Anual Asoc. Venez. Avance de la Ciencia, Barquisimeto.

González, C., J. Goicochea, and F. Perozo. 1979b. Peso al nacimiento y tasa de crecimiento en corderos West Africa. Idem. Barquisimeto.

Hernández, I. 1977. Peso vivo, peso en canal y rendimiento en canal en ovinos y caprinos sacrificandos en el FIBCA (Estado Zulia). Trabajo de ascenso. Copia multigraf. 67 pp. Fac. Agronomía, Univ. Zulia, Maracaibo.

REPRODUCTION AND GROWTH OF HAIR SHEEP IN AN EXPERIMENTAL FLOCK IN VENEZUELA

Andres Martinez, *Winrock International, U.S.A.*

INTRODUCTION

The sheep industry in Venezuela is in the developmental stage. During the past two decades, the national herd has increased rapidly, reaching a total population of about 294,000 in 1978 (USDA/FAS 1980). Reverón et al. (1976) note that although sheep meat represents only 0.1% (500 metric tons annually) of the national red meat consumption, the potential for growth is favorable due to: (1) wide acceptance of sheep meat among the population; (2) producer interest in commercial production; and (3) adaptability of various breeds, particularly hair sheep breeds, to the arid and semiarid zones of the country.

Prior to 1960, Venezuela had little research in sheep production; since then, however, the Centro Nacional de Investigaciones Agropecuarias (CENIAP) has supported the improvement of the sheep production sector, primarily through its Instituto de Investigaciones Zootécnicas. Research efforts at the Institute have focused mainly on the characterization of local and imported breeds and their crosses. The following observations are based on results obtained at the facilities of the Institute located in Maracay.

ENVIRONMENT AND FLOCK MANAGEMENT

Climate

The city of Maracay is located in north-central Venezuela, approximately 90 km west of Caracas. The climate at the experiment station is characterized

This review is based on work by A. Reverón, G. Mazzarri, C. Fuenmayor, and V. Bodisco, Instituto de Investigaciones Zootécnicas, Centro Nacional de Investigaciones Agropecuarias.

by distinct wet-dry seasons and little fluctuation in temperature. Annual precipitation is close to 1 m, with the rainy season extending from May through November. Ambient temperature varies from 22 to 25° C and relative humidity from 65 to 80%. The lower humidity values correspond to the dry season.

Animal Management

General management of the station's flock is summarized in Table 2.5.1, from descriptions reported by Bodisco et al. (1973). The sheep are maintained throughout the year on pasture comprised of Star (*Cynodon plectostachium*) or Pangola (*Digitaria decumbens*) grasses, or both. Chopped elephant grass (*Pennisetum purpureum*) is provided ad libitum during the dry season to supplement the low nutritive levels of Star and Pangola pastures. Animals are supplemented with a 14% protein concentrate during the breeding season and at lambing, as indicated in Table 2.5.1.

The breeding season extends from August through October. All matings are natural; rams are placed in pastures containing from 10 to 50 ewes at the rate of one ram for every 10 ewes. After lambing, the ewes and their lambs spend the first 30 days under a shed and the subsequent 60 days on pasture. In cases of triple births, one of the lambs is initially fed 800 g of cow's milk daily. All lambs are normally weaned at 90 days of age.

Animal Resources

The station's sheep flock consists of four breeds—Criollo, Barbados Blackbelly, West African, and Blackhead Persian. The latter three breeds were imported in the early 1960s from the Caribbean islands of Barbados and Trinidad/Tobago. Reverón et al. (1976) describe the Criollo breed as follows:

> These sheep have a small head with a wide woolly forehead, a wide white face (with or without pigmentation) and short ears (with or without dark spots) covered with fine hair. The neck is short and forms a straight line with the cylindrical trunk. The back is straight or swayed and the rump is drooping. The tail is long and is inserted low in the rump. The legs are long, well-developed, and covered with wool for two-thirds of their length. The skin is pale pink, fine, and usually without the wrinkles that may be found on the necks of some animals. Some males have horns. Generally, the fleece extends throughout the body and down to the knees, but the wool is of very low quality.

Criollo sheep are descendants of animals brought to the Americas from the Iberian Peninsula. Harsh environmental conditions and several centuries of indiscriminate matings have resulted in small, slow-maturing but well-adapted animals. Criollo sheep are by far the most numerous in the country.

TABLE 2.5.1.   MANAGEMENT OF EWE FLOCK

Activity	July	Aug	Sept	Oct	Nov	Dec	Jan	Feb	Mar	Apr	May	June
Reproduction												
Breeding season			———	———								
Lambing							——	— —	— —			
								30 days				
Nutrition												
Pasture	———	———										
Chopped grass		21 days										
Supplement (14% protein concentrate)		— 90 days										
Mature ewes: 0.5 kg/head/day	30 days											
Ewe lambs: 0.25 kg/head/day												
Rams: 0.5 kg/head/day												
Climatic data (1975 and 1976)												
Rainfall in mm	208	167	117	133	104	43	1	13	12	26	78	113
Mean ambient temperature, °C	24	24	24	24	24	23	23	23	25	25	26	25
Relative humidity, %	76	78	77	78	78	68	70	67	67	68	70	73

TABLE 2.5.2.   LITTER SIZE OF VARIOUS BREEDS OF SHEEP

Breed	No. of litters	Percent ewes having			Litter size	
		Singles	Twins	Triplets	x̄	SD
West African	277	58.8	39.4	1.8	1.43	.53
Barbados Blackbelly	195	61.0	33.3	5.6	1.45	.60
Criollo	227	87.7	11.5	0.9	1.13	.36
Blackhead Persian	83	96.4	3.6	0	1.04	.19

Adapted from Bodisco et al. (1973).

MEASURES OF PERFORMANCE

Four parameters—reproduction, size, growth rates, and mortality rates—are used in this discussion to compare the performance of hair sheep and their crosses.

Reproduction

Age at First Breeding. Ewes are traditionally bred first at one year of age (Bodisco et al. 1973), largely because of the established practice of lambing during the dry season. Determination of the most appropriate age and weight for first breeding has been hampered by lack of data. Reverón et al. (1979), however, demonstrated that it is possible to breed crossbred (Criollo × West African) ewe lambs at 10 months of age without decreasing lamb birth weights and weaning weights significantly, as compared to weights of lambs born to ewes bred at 12 months of age.

Litter Size. Table 2.5.2 shows the percentages of ewes having single and multiple births for various breeds. There is a tendency for West African ewes to twin more often than do Barbados Blackbelly ewes (39 vs 33%). However, the latter have more triplets. The incidence of multiple births is very low (less than 4%) for Blackhead Persian ewes and intermediate (12.4%) for Criollo ewes. Mazzarri and Fuenmayor (1979) report somewhat similar multiple-birth levels for West African and Barbados Blackbelly (38 and 49%, respectively), but report a much higher rate (23%) for Criollo ewes than did Bodisco et al. (1973).

In addition to breed differences, season of the year also affects reproductive performance. Overall, multiple births for the station's flock are approximately two times greater during the rainy season than in the dry season, according to Mazzarri and Fuenmayor (1979). Trials conducted by Mazzarri et al. (1976) indicate that supplemented ewes (forage ad libitum plus concentrate) had a significantly greater number of lambs per ewe and a

TABLE 2.5.3.   FERTILITY OF SHEEP FOLLOWING ESTRUS
                 SYNCHRONIZATION

Breed	Animals treated, number	Ewes pregnant at:		Parturi- tions	Litter size
		First estrus	Second estrus		
West African	15	11	4	15	1.66
Barbados Blackbelly	12	11	1	12	1.75
Criollo	15	10	4	14	1.40
Total	42	32	9	41	1.63
%	–	76.2	21.4	96.6	

Source: Mazzarri et al. (1973).

larger percentage of multiple births than did ewes receiving only forage ad libitum, or ewes restricted to 50% of ad libitum forage.

Duration of Estrus and Gestation. Using the intravaginal sponge technique to synchronize estrus in one-year-old ewes, Mazzarri et al. (1976) observed the following estrous and gestation periods by breed:

	West African	Barbados Blackbelly	Criollo
Duration of estrus, hours	38.6	34.3	24.5
Duration of gestation, days	148.6	150.2	147.0

There were no highly significant differences ($P < 0.05$) in the length of estrus between West African and Barbados Blackbelly ewes; however, Criollo ewes showed significantly shorter estrous length than the other two breeds. Supplementation of a basic roughage diet with a high-protein concentrate had no effect on the length of estrus or gestation.

Estrus Synchronization. Hair sheep in tropical environments generally breed throughout the year. Lactational anestrus, however, limits the opportunity for two gestations in one year. Tests on synchronization of estrus and its subsequent effect on fertility were conducted by Mazzarri et al. (1973). All treated ewes showed estrus within 24 to 72 hours after removal of intravaginal sponges; 93% of the control ewes showed estrus during a 30-day period. Lambing-rate was 97.6 and 85.3% for treated and control ewes, respectively. The fertility of synchronized animals (Table 2.5.3) was high, and litter size somewhat higher than normally obtained in untreated ewes (Table 2.5.2).

Fuenmayor et al. (1978) further demonstrated that estrus synchronization could reduce the age at first breeding from age 12 months to age 10 months

TABLE 2.5.4.   SUMMARY OF LAMB BIRTH WEIGHTS BY BREED AND
                   TYPE OF BIRTH FOR THE PERIOD 1968-1972, kg

Breed	Type of birth			Weighted average
	Single	Twin	Triplet	
West African	3.12	2.57	1.90	2.78
Barbados Blackbelly	2.73	2.49	1.99	2.54
Criollo	2.87	2.44	1.58	2.76
Weighted Average	2.92	2.53	1.92	2.70
Blackhead Persian	2.50	2.08	-	2.49

Source: Bodisco et al. (1973).

in tropical sheep, without adverse effects on fertility and birth weights. Although data on estrus synchronization appear encouraging, the economic feasibility and practicality of the technique have not been evaluated under prevailing field conditions.

Size and Growth

Animal productivity is known to be a function of heredity and environment, and of possible interactions between them. Generally, most animals indigenous to the tropics are less productive than their counterparts in temperate climates because of harsher environments and minimal selection pressure for productive traits.

This section deals selectively with the effects of a few sources of variation such as breed, sex, and type of birth (single vs multiple) on size and growth of tropical sheep. Three chronological reports of research activities at the station are discussed: a report by Bodisco et al. (1973) summarizes data for the years 1968 through 1972; two papers by Reverón et al. (1978a; 1978b), summarize data from 1972 through 1977.

Birth Weights. There is good evidence of a positive correlation between birth weights and subsequent weights. For this reason, considerable attention has been given to birth weights as an important factor in improving the productivity of tropical sheep in Venezuela (Bodisco et al. 1973; Reverón et al. 1978a; Fuenmayor et al. 1978).

Average lamb birth weights for purebred breeds ranged from 2.49 to 2.78 kg (Table 2.5.4) during the period 1968–1972. West African lambs had the highest weight followed in descending order by Criollo, Barbados Blackbelly, and Blackhead Persian lambs.

Type of birth (single vs multiple) consistently has been shown to affect birth weights. Weighted averages of 2.92 kg for singles, 2.53 kg for twins, and 1.92 kg for triplets are reported in Table 2.5.4. Weights of twins were 86.7% of the weights of singles. Triplet weights were 65.8% of single

TABLE 2.5.5.   LAMB BIRTH WEIGHTS AS AFFECTED BY TYPE OF BIRTH

| Source | Birth weight, kg[a] | | |
	Single	Multiple	$\frac{Multiple}{Single}$ %
Fuenmayor et al. (1978)	2.95	2.41	81.7
Reverón et al. (1979)	3.00	2.50	83.3
Reverón et al. (1978a)	2.76	2.36	85.5
Reverón et al. (1978b)	2.98	2.52	84.6

[a] Differences between single and multiple weights were statistically significant in all cases.

weights and 75.9% of twin weights. More recent data on birth weights, including those from crosses of Criollo ewes with the three imported breeds, also indicate a significant difference in birth weights between single and multiple births (Table 2.5.5). Average birth weights varied between 2.75 and 3.00 kg for single births and from 2.35 to 2.50 kg for multiple births. The birth weight of multiples represented about 82 to 86% of the birth weight of singles. Breed and type of birth interactions also affected birth weights (Bodisco et al. 1973).

Sex of the newborn lamb had no significant effect on birth weights analyzed by Bodisco et al. (1973), Fuenmayor et al. (1978), Reverón et al. (1978a), and Reverón et al. (1979). Conversely, newborn male lambs from crosses of Criollo ewes with Blackhead Persian, Barbados Blackbelly, and West African rams weighed significantly more (2.81 vs 2.69 kg) than their counterpart females (Reverón et al. 1978b).

The age of the ewes showed a positive and significant effect on lamb birth weight in the five-year data analysis of Bodisco et al. (1973). This effect was not apparent in ewe lambs first bred at 10 to 12 months of age (Fuenmayor et al. 1978; Reverón et al. 1979). Synchronization of estrus had no effect on lamb birth weight.

Weaning Weights. Lambs are normally weaned at 90 days of age. Weights at this age for purebreds are summarized in Table 2.5.6. There were only small differences in weights among West African, Barbados Blackbelly, and Criollo lambs. Blackhead Persian lambs weighed approximately 2 kg less than did lambs from the other three breeds.

As was the case with birth weights, the type of birth significantly affected weaning weights of purebred lambs (Table 2.5.6). Twins weighed 86% of the weight of singles. Triplets weighed 79% of singles' weights and 92% of

TABLE 2.5.6.   SUMMARY OF LAMB WEANING WEIGHTS (90 DAYS)
                BY BREED AND TYPE OF BIRTH FOR THE PERIOD
                1968-1972, kg [a]

Breed	Type of birth Single	Twin	Triplet	Weighted average
West African	14.0	11.4	9.9	12.5
Barbados Blackbelly	12.9	11.8	10.5	12.1
Criollo	12.5	10.3	11.1	12.1
Weighted Average	13.1	11.3	10.4	12.2
Blackhead Persian	10.3	9.1	-	10.2

Source: Bodisco et al. (1973).

TABLE 2.5.7.   LAMB WEANING WEIGHTS (90 DAYS) AS AFFECTED
                BY TYPE OF BIRTH

Source	Weaning weight, kg[a] Single	Multiple	Multiple Single %
Reverón et al. 1978a[a]	14.7	13.8	93.7
Reverón et al. 1978b[b]	14.0	12.3	88.1
Reverón et al. 1979 [c]	16.6	14.7	88.6

[a] Average weights of purebred and crossbred lambs

[b] Average weights of crossbred lambs - Criollo with West African,
     Blackhead Persian or Barbados Blackbelly

[c] Average weights of crossbred lambs - Criollo with West African

twins' weights. Data for mostly crossbred lambs (Table 2.5.7) also indicate
differences in weaning weights between single and multiple birth lambs.
This difference is significant only in the report by Reverón et al. (1978b).
Weaning weights for multiple births presented in Table 2.5.7 are 88 and
94% of the weights of single births.

Other sources of variation affecting weaning weights were year of birth
and an interaction between year and breed (Bodisco et al. 1973). None of
the papers reviewed reported that the sexes differed in weaning weights
(Bodisco et al. 1973; Reverón et al. 1978a, 1978b, and 1979; Fuenmayor et
al. 1978). The significant difference in birth weights between males and

TABLE 2.5.8.   LAMB WEIGHTS AT SIX MONTHS OF AGE BY BREED AND TYPE OF BIRTH, kg

| Breed | Type of birth | | |
	Single	Twin	Triplet
West African	19.5	16.9	14.3
Barbados Blackbelly	22.2	19.6	17.2
Criollo	16.7	20.0	14.0
Weighted average	18.9	17.8	16.1
Blackhead Persian	15.2	14.9	-

Source: Bodisco et al. (1973).

TABLE 2.5.9.   LAMB WEIGHTS AT SIX MONTHS OF AGE, kg

| Source | Type of birth | | Multiple Single % |
	Single	Multiple	
Reverón et al. (1978a)	21.5	20.1	93.4
Reverón et al. (1978b)	18.6	17.2	92.4

females found by Reverón et al. (1978b) was not evident in the comparison of weaning weights.

Six-Month Weights. Average six-month weights of purebred lambs are shown in Table 2.5.8. Barbados Blackbelly lambs were the heaviest, followed by West African lambs at 75% of Barbados Blackbelly lamb weights, Criollo lambs at 67%, and Blackhead Persian lambs at 62%. The effect of type of birth on body weight had largely disappeared by the age of six months, although there was a tendency for triplets and twins to be lighter than singles. Triplets weighed 85% of the weight of singles, and twins weighed 94% of singles' weights, but these differences were not statistically significant. Conversely, Reverón et al. (1978a; 1978b) found that type of birth had a significant effect on six-month weights. In these studies, the multiple-birth lambs weighed 92 to 93% of the weight of singles (Table 2.5.9).

Other sources of variation affecting six-month weights of purebred lambs included interactions between breeds and years and between sex and years (Bodisco et al. 1973). Also, Reverón et al. (1978a) reported significantly higher six-month weights (21.0 vs 19.8 kg) for purebred lambs than for crossbred lambs. Sex of the lamb had no effect on six-month weights (Reverón et al. 1978a and 1978b).

TABLE 2.5.10.   DAILY GAINS OF LAMBS DURING SUCKLING
                     (0-90 DAYS OF AGE), g

| | Type of birth | | |
Breed	Single	Twin	Triplet
West African	121	97	89
Barbados Blackbelly	112	103	92
Criollo	86	86	102
Weighted average	113	97	93

Source: Bodisco et al. (1973).

TABLE 2.5.11.   DAILY GAINS OF LAMBS FROM 90 to 180 DAYS
                     OF AGE, g

| | Type of birth | | | Weighted |
Breed	Single	Twin	Triplet	average
West African	50	48	33	49
Barbados Blackbelly	84	72	69	77
Criollo	30	32	22	30
Weighted average	48	53	55	50

Source: Bodisco et al. (1973).

Yearling Weights. Very little information is available on comparative weights of tropical sheep at 12 months of age. Mazzarri et al. (1976) report an average weight of 23.5 kg for 12-month-old acyclical West African, Barbados Blackbelly, and Criollo ewes, while Fuenmayor et al. (1978) report average weights of 22 and 24 kg for 10- and 12-month-old ewe lambs, respectively.

Growth Rate

Average daily increases in body weight of purebred lambs during the 90-day suckling stage, the next 90 days after weaning, and overall for the first 180 days are shown in Tables 2.5.10, 2.5.11, and 2.5.12, respectively. During the suckling period lambs averaged a daily weight gain of 105 g (Table 2.5.10). No significant differences were found between breeds. Type of birth affected gain significantly, with twins and triplets gaining less than singles (97 and 93 vs 113 g).

During the 90-day period after weaning, the average daily gain dropped to less than one-half (50 g) that of the preweaning daily gain. Breed significantly affected the weighted average daily gain, although type of birth had no effect (Table 2.5.11). Barbados Blackbelly lambs had the fastest

TABLE 2.5.12.   DAILY GAINS OF LAMBS DURING THE FIRST 180 DAYS, g

Breed	Type of birth			Weighted average
	Single	Twin	Triplet	
West African	84	71	60	77
Barbados Blackbelly	99	87	75	90
Criollo	68	60	57	66
Weighted average	80	74	70	77

Source: Bodisco et al. (1973).

gains—2.6 and 1.6 times the daily gain of Criollo and West African lambs, respectively. Triplets tended to gain more weight than twins and the latter more than singles. Bodisco et al. (1973) point out that more attention should be given to the overall nutrition of weaned lambs to prevent heavy drops in daily gains during the postweaning period.

Overall daily weight gain averages for the first 180 days are presented in Table 2.5.12. Breed and type of birth significantly affected daily gains. Barbados Blackbelly lambs had the largest average daily gains (90 g).

Gains for West African lambs were 86% of the gains of Barbados Blackbelly lambs; those for Criollo lambs were 73%. Single lambs gained at a significantly greater rate than did multiple-birth lambs. Twin lambs gained 4 g daily more than did triplets, although this difference was not statistically significant.

Tables 2.5.10, 2.5.11, and 2.5.12 indicate a well-defined advantage of the imported breeds over the Criollo sheep. Of the imported breeds, the Barbados Blackbelly had a faster growth rate than did the West African during the first 180 days.

Mortality Rates

Mortality rates of purebred lambs born at the experiment station from 1968–1972 have been reported by Bodisco et al. (1973). Table 2.5.13 shows an overall mortality rate of approximately 28% from birth to 6 months of age, with no significant differences between breeds. There is a tendency, however, for death rates to be higher among Blackhead Persian and Barbados Blackbelly lambs than among West African and Criollo lambs.

The particular period (0–90 vs 90–180 days) had a definite effect on mortality (Table 2.5.13). Approximately 4% more lambs died during the preweaning period than died during the postweaning period. Overall, about 15% more lambs from triplet births died than did lambs from single births (Table 2.5.14). Mortality rate of Barbados Blackbelly lambs differed little among the three types of births; the other two breeds had a substantial, pro-

TABLE 2.5.13.   MORTALITY RATE OF LAMBS FROM BIRTH TO 6
                  MONTHS BY BREED, %

Breed	Period, days		
	0-90	90-180	Total
West African	15.8	9.9	25.7
Barbados Blackbelly	19.1	15.4	34.5
Criollo	10.8	10.0	20.8
Blackhead Persian	21.0	13.6	34.6
Average	15.9	11.7	27.6

Source: Bodisco et al. (1973).

TABLE 2.5.14.   EFFECT OF BREED AND TYPE OF BIRTH
                  ON LAMB MORTALITY RATES

Breed	Type of birth			
	Single	Twin	Triplet	Total
West African	16.3	32.1	38.5	25.7
Barbados Blackbelly	33.6	35.2	35.5	34.6
Criollo	18.9	26.5	40.0	20.8
Average	21.6	32.4	36.7	27.0

Source: Bodisco et al. (1973).

gressive increase in death rates—from singles to twins to triplets. Mortality
rates of West African and Criollo triplets were 2.36 and 2.11 times greater
than were those for their counterpart single-birth lambs.

SUMMARY

This paper reviews some results obtained during the past decade at the Institute of Animal Research in Maracay (Venezuela) on the characterization
of tropical sheep breeds and some of their crosses. Animal resources include
the native or Criollo breed of sheep and three breeds imported in the early
1960s—Barbados Blackbelly, West African, and Blackhead Persian.

Primary measures of performance included reproduction, size and
growth, growth rates, and mortality rates. Among the reproductive
characteristics, litter size was found to be an important variable. Number
of lambs per parturition averaged 1.45 for Barbados Blackbelly, 1.43 for
West African, 1.13 for Criollo, and 1.04 for Blackhead Persian sheep.

Average birth weights ranged from 2.78 to 2.49 kg for the four breeds,
with West African and Criollo lambs having the heaviest weights and
Blackhead Persian the lowest. Type of birth influenced birth weights. Twins

averaged 86.7% of the birth weight of singles, and triplets averaged 65.8% of the birth weight of singles.

Average daily gain during the suckling period (first 90 days) varied in different trials from 105 to 142 g. Gains during the first six months ranged from 77 to 100 g per day. Type of birth, breed, and sex of the lamb affected rate of gain during this period. Of the purebred sheep, Barbados Blackbelly had the highest daily gains (90 g) to six months of age, followed by West African (77 g), and Criollo (66 g). Single lambs gained about 10 g daily more during this period than did their counterpart triplets.

Mortality rates averaged 28% from birth to six months of age, with greater mortality during the first three months (15.9 vs 11.7%) and a tendency for higher death rates for multiple births as compared to singles.

REFERENCES

Bodisco, V., C. M. Duque, and A. Valle S. 1973. Comportamiento productivo de ovinos tropicales en el período 1968–1972. *Agronomía Tropical* 23(6):517–540.

Fuenmayor, C., G. Mazzarri, V. Bodisco, and A. E. Reverón. 1978. Effectos de la sincronización del estro sobre la fertilidad de corderas tropicales. *Memoria, Asociación Latino Americana de Producción Animal.* 13:164 (Abstract).

Mazzarri, G. and C. Fuenmayor. 1979. Comportamiento reproductivo de las ovejas tropicales. Unpublished data. Centro Nacional de Investigaciones Agropecuarias, Ministerio de Agricultura y Cría, Maracay, Venezuela.

Mazzarri, G., C. Fuenmayor, and C. F. Chicco, 1976. Efecto de diferentes niveles alimenticios sobre comportamiento reproductivo de ovejas tropicales. *Agronomía Tropical.* 26:205.

Mazzarri, G., C. Fuenmayor, and C. M. Duque. 1973. Control del ciclo estral mediante el uso de esponjas vaginales impregnadas en acetato de fluorogestona en ovejas. *Agronomía Tropical.* 23:315–321.

Reverón, A. E., V. Bodisco, M. Arriojas, C. F. Chicco, and H. Quintana. 1978a. Crecimiento de corderos tropicales. IV Conferencia Mundial de Producción Animal, Buenos Aires, Argentina.

Reverón, A. E., V. Bodisco, M. Arriojas, and H. Quintana. 1978b. Comportamiento productivo en dos generaciones filiales de ovejas tropicales. IV Conferencia Mundial de Producción Animal, Buenos Aires, Argentina.

Reverón, A. E., V. Bodisco, G. Mazzarri, M. Arriojas, and C. Fuenmayor. 1979. Effecto de la edad al primer servicio sobre el crecimiento de corderos tropicales. A ser presentado en la VII Reunion del ALPA en Panama.

Reverón, A. E., G. Mazzarri, and C. Fuenmayor. 1976. *Ovejas tropicales productoras de carne.* Maracay, Venezuela: Centro Nacional de Investigaciones Agropecuarias.

USDA/FAS. 1980. Livestock and meat. Foreign Agriculture Circular FLM 2-80. Washington, D. C.

averaged 66.7% of the birth weight of singles, and triplets averaged 65% of the birth weight of singles.

Average daily gain during the suckling period (first 90 days) varied in different trials from 105 to 142 g. Gains during the first six months ranged from 77 to 100 g per day. Type of birth, breed, and sex of the lamb affected rate of gain during this period. Of the purebred sheep, Barbados Blackbelly had the highest daily gains (90 g) to six months of age, followed by West African (72 g), and criollo (66 g). Single lambs gained about 10 g daily more during this period than did their counterpart triplets.

Mortality rates averaged 28% from birth to six months of age, with greater mortality during the first three months (15.9 vs 11.8%) and a tendency for higher death rates for multiple births as compared to singles.

REFERENCES

Bodisco, V.; C. M. Duque, and A. Valle S. 1973. Comportamiento productivo de ovinos tropicales en el período 1968-1972. Agronomia Tropical 20(6):517-540.

Fuenmayor, C.; C. Mazzarri, V. Bodisco, and A. E. Reverón. 1978. Efectos de la suplementación del zinc sobre la fertilidad, te (corderas tropicales). Memoria, Asociación Latino Americana de Producción Animal. 13:164 (Abstract).

Mazzarri, G. and C. Fuenmayor. 1975. Comportamiento reproductivo de las ovejas tropicales. (Unpublished data). Centro Nacional de investigaciones agropecuarias, Ministerio de Agricultura y Cria, Maracay, Venezuela.

Mazzarri, G., C. Fuenmayor, and C. F. Chicco. 1976. Efecto de diferentes niveles de alimentación sobre comportamiento reproductivo de ovejas tropicales. Agronomia Tropical 26:205.

Mazzarri, G., C. Fuenmayor, and C. M. Duque. 1973. Control del celo e inseminación artificial en ovejas mediante impregnadas en acetato de fluorogestona en trópico. Agronomia Tropical 23:315-323.

Reverón, A. E., V. Bellora, M. Arreaza, C. F. Chicco, and E. Quintana. 1978. Crecimiento de corderas tropicales. IV Conferencia Mundial de Producción Animal. Buenos Aires, Argentina.

Reverón, A. E., V. Bodisco, V. Arreaza, M. Arreaza, and C. Fuenmayor. 1978. Efecto de la edad al primer servicio sobre la reproducción de corderas tropicales. A ser presentado en la VII Reunión del ALPA en Panamá.

Reverón, A. E., V. Mazzarri, and C. Fuenmayor. 1976. Datos no publicados. Instituto de Producción Animal, Maracay, Venezuela. Centro Nacional de Investigaciones Agropecuarias.

USDA-FAS. 1980. Livestock and meat. Foreign Agriculture Circular FLM 2-80. Washington, D. C.

PERFORMANCE OF BARBADOS BLACKBELLY SHEEP AND THEIR CROSSES AT THE EBINI STATION, GUYANA

G. Nurse, N. Cumberbatch, and P. McKenzie
Ministry of Agriculture, Guyana

BACKGROUND

The Guyana Ministry of Agriculture is developing technology to expand sheep production in Guyana. Sheep research was initiated at three locations during the 1970s with the principal focus at Ebini, Berbice River. Devendra (1975) has provided some general information on sheep production in Guyana, and data supplied by the Guyana Ministry of Agriculture from two stations (Moblissa and Ebini) have been reported by Devendra (1977).

The Ebini Station is located at approximately 6° N, 58° W and has a mean annual precipitation of 2,160 mm. There are two main seasons for precipitation: mid-April to late August and November to late January. Mean annual temperature is approximately 27° C, with a maximum variation in monthly means of about 1° C. Relative humidity averages about 78%.

In general, the soils of the area are very low in nutrient content, requiring lime and fertilizer to support healthy plant growth. Forage species include *Andropogon leuchostachyus, Digitaria* spp. (*decumbens, pentzii* and *setivalva*) and *Brachiaria* spp. For normal performance, livestock may require supplements of phosphorus, calcium, magnesium, molybdenum, copper, cobalt, manganese, boron, and possibly other elements. "Hind leg weakness," probably due to plant poisoning, is a frequent problem, especially during the longer dry season.

Sheep of the Barbados Blackbelly and Creole (native) breeds have been introduced to the Ebini Station, along with some goats. In general, the sheep are performing better than the goats, though mortality rates of young animals are a problem in both species.

Traditionally, rams ran with the ewes year-round. Lambings occurred throughout the year but with some seasonal peaks related to seasonal forage growth; lambing interval ranged from 180 to 440 days, with a distinct peak at 210 days (McPherson 1976).

PERFORMANCE AT EBINI STATION

By 1978, breeding management had been imposed so that lambing occurred during restricted periods. Table 2.6.1 shows the performance of 41 Blackbelly ewes and 22 ewes of Creole (mixed) breeding, which lambed in August and in November after being bred to Blackbelly rams.

The two groups of ewes differed in weight by more than 6 kg, with the Blackbelly ewes being heavier. The weight of 33 kg is markedly higher than the 23 kg cited by Devendra (1975). Mean litter size was much higher for the Blackbelly ewes, but lamb mortality also was much higher in this group, resulting in little difference in lambs raised per ewe lambing (0.95 vs 0.91).

Lambs were weaned at 100 to 132 days, and 120-day weights were calculated by linear interpolation between (or extrapolation from) birth and weaning weights. Table 2.6.2 summarizes birth and 120-day weights.

Birth weights of lambs from the smaller dams of mixed (Creole) breeding were actually slightly higher than those of straight Blackbelly dams. Average birth weights of single lambs were 10.4% of the weight of mature dams of the Creole group; 7.7% of the weight of Blackbelly dams. In both groups, weights of lambs of multiple births were only slightly less than those of single-born lambs. All lambs were relatively light at weaning; gain from birth to weaning averaged about 90 g per day.

The survival and growth of the crossbred group of lambs suggest that crossing of Barbados Blackbelly rams with ewes of Creole breeding may be a useful mating plan; performance in terms of litter size, lamb survival, and lamb growth should be of interest when the females from this cross are used for breeding.

SUMMARY

Barbados Blackbelly ewes, 2 years old and over, had a mean body weight at mating of 32.7 kg and a mean litter size at birth of 1.70. Lamb survival was 57%, and the lambs averaged 13 to 14 kg at 4 months of age. A group of ewes of Creole breeding mated to Blackbelly rams were smaller than the Blackbelly ewes and only 10% had twins, but lamb survival was 83% and weights of lambs at birth and weaning were similar to those of the straight Blackbelly lambs.

TABLE 2.6.1. LITTER SIZE AND MATING WEIGHT OF BLACKBELLY AND MIXED BREED EWES, EBINI STATION, 1978

Ewe group	Ewe age	No. ewes	Weight, kg		No. ewes with litters of:			Mean lambs/ewe		% lamb survival
			x̄	SD	1	2	3	Born	Weaned	
Blackbelly	1 yr	8	24.3	5.2	4	4	–	1.50	0.75	
	2 yr	33	32.7	6.3	12	19	2	1.70	1.00	
	All	41			16	23	2	1.66	.95	57
Mixed[a]	1 yr	2	22.9	4.2	2	–	–	1.00	0.50	
	2 yr	20	26.4	5.0	18	2	–	1.10	0.95	
	All	22			20	2	–	1.09	.91	83

a Creole or crossbred

TABLE 2.6.2.   BIRTH AND 120-DAY WEIGHTS OF PURE AND CROSSBRED BARBADOS BLACKBELLY (BB) LAMBS BY TYPE OF BIRTH, kg

Sire	Dam	Litter size	Sex	Birth weight				120-day weight		
				No.	$\bar{x}$	SD	Ratio[a]	No.	$\bar{x}$	SD
BB	BB	1	M	6	2.37	.59		3	13.7	3.5
			F	11	2.69	.76		6	13.8	2.4
		2	M	22	2.12	.58	.88	10	13.4	2.9
			F	24	2.34	.51		14	12.3	1.7
		3	M	4	2.10	.08	.87	–		
			F	2	2.30	.28		–		
BB	Mixed	1	M	14	2.94	.64		11	15.5	2.9
			F	6	2.57	.74		6	12.9	3.1
		2	M	2	2.44	.08	.93	2	10.1	.4
			F	2	2.67	.08		1	12.4	–

[a] Ratio of birth weight of multiples to birth weight of singles (based on unweighted average of means for males and females).

REFERENCES

Devendra, C. 1975. Sheep and goat production in Guyana. *Z. Tierzucht. Zuchtgsbiol.* 92:305–317.

Devendra, C. 1977. Sheep of the West Indies. *World Review of Animal Production* 13(1):31–38.

McPherson, V. O. 1976. A review of activities of the sheep and goat sector 1973–1976. June 1976. Mimeo. 19 pp. Ministry of Agriculture, Ebini, Guyana.

HAIR SHEEP PERFORMANCE IN BRAZIL

**E.A.P. de Figueiredo, E. R. de Oliveira,
C. Bellaver, and A. A. Simplício**
*Centro Nacional de Pesquisa de Caprinos,
Empresa Brasileira de Pesquisa Agropecuária, Brasil*

INTRODUCTION

Hair sheep production in Brazil is centered in the northeast region bounded by latitudes 1° N and 18° 30′ S and longitudes 34° 30′ W and 48° 20′ W. The northeast region contains 1,640,000 km² (approximately one-fifth of the total Brazilian surface) and includes 9 states (Maranhao, Piaui, Ceará, Rio Grande do Norte, Paraiba, Pernambuco, Alagoas, Sergipe, and Bahia) of the 22 states in Brazil (Andrade 1977).

According to the *Anuário Estatístico do Brasil* (1977), regional sheep population is estimated at 5,289,912 head (distribution by state is shown in Table 2.7.1). Although this census did not identify breed or type, it has been estimated that 85 to 90% of the sheep flocks in this region are hair sheep or have hair sheep ancestors. Approximately 10% of the sheep are established breeds, such as Morada Nova, Santa Inês, and Brazilian Somali.

ORIGIN AND CHARACTERISTICS OF HAIR SHEEP IN BRAZIL

Hair sheep in Brazil include the *Pelo do Boi* (hair of ox), the generic term for the population of grades and true breeds, Morada Nova, and Santa Inês; *Rabo Gordo* (fat rump) primarily Brazilian Somali Blackhead; and *Rabo Largo* and *La Grosseira* (fat tail and grades) that show traces of the Italian Bergamasca breed; as well as local unimproved types.

Morada Nova

The origin of this breed is the subject of controversy. Domingues (1954) proposed that this breed was directly descended from the Portuguese medium-wool breed, Bordaleiro, which was introduced into Brazil at the

TABLE 2.7.1.   SHEEP POPULATION IN THE NORTHEAST REGION OF BRAZIL

State	Head
Maranhao	119,690
Piaui	788,887
Ceará	1,065,534
Rio Grande Do Norte	272,260
Paraiba	359,775
Pernambuco	476,963
Alagoas	127,947
Sergipe	109,069
Bahia	1,969,787
TOTAL	5,289,912

Source: Anuário Estatístico do Brasil (1977).

time of Portuguese colonization. Mason (1980) has proposed that the Morada Nova is descended principally from West African sheep, probably brought to Brazil on slave ships. It also is possible that the breed is descended from crosses between sheep introduced from Portugal and Africa.

The Morada Nova is selected for the absence of wool. Color is usually red, with a white tail tip and black hooves, but white is also accepted. Animals that do not meet these requirements are culled. Breed standards set by the Associação Brasileira de Criadores de Ovinos (ABCO 1977) include the following: no horns (rudimentary horns may be allowed for rams); pointed shell-shaped ears, 9 cm in length; long head, subconvex profile; short, slightly sloping rump; thin tail; red or white hair color; dark skins, mucous membranes, and hooves; and short, thick, shiny hair. The Morada Nova is considered a dual-purpose animal for meat and leather production; its hide has good commercial value. Disqualifying characteristics include: wool; no white tip on tail; unpigmented skin and mucous membrane; horns; large or pendent ears; beard or mane; short, fat tail; spots of any color; and genital defects.

Santa Inês

The Santa Inês is large. Females are highly fertile, frequently twinning. Milk production is good. A good environment, especially high levels of nutrition, is required to support this large, highly productive breed.

Colors include white, red, and black; spotting is acceptable. Ears are long and pendent. The breed is thought to have been developed from crosses between the large (80- to 90-kg adult-ram weight), white, coarse-wooled Italian Bergamasca breed and the Brazilian hair sheep. Selection of the Santa Inês has been principally for size and absence of wool. Standards for the breed include (ABCO 1977): polled; medium-length, thin tail; heavy bone; and black or white hooves. Disqualifying characteristics include: ultra-convex profile, small size, skeletal defects, excess fat deposits, and lack of skin pigmentation for red, spotted, or black animals (white skin is acceptable if hair is white).

Rabo Largo

The Rabo Largo is a horned sheep having a long, fat tail; colors are white, spotted, or white with colored head. According to Mendonça (1951), they were introduced from South Africa. The original imports were large, short-haired, woolless, and had very large fat tails (approximately 20 cm wide and S-shaped). These imports were crossed with smaller, wooled, thin-tailed sheep in Brazil. Present-day Rabo Largo sheep are horned, partially wooled, and have a long, slightly fat tail.

Brazilian Somali

The Brazilian Somali is black- or brown-headed, tailless, and fat-rumped. The earliest introduction to Brazil was made by a Rio de Janeiro producer, Pinheiro Jr., in 1939. Some of these sheep carry a small amount of wool, but most individuals are woolless.

There are few registered sheep for these breeds. Candidates for registration are inspected, but only a minority are registered. In 1978, the numbers for each breed inspected and the numbers registered (in parentheses) in the northeast region were: Morada Nova—17 (8) males, 361 (60) females; Santa Inês—45 (9) males; 456 (67) females; Bergamasca—2 (2) males, 6 (4) females; Brazilian Somali—4 (2) males, 10 (2) females.

PRODUCTION ENVIRONMENT

Climate

Most sheep are found in the areas of 600 to 1000 mm average annual rainfall. However, annual rainfall is highly irregular, ranging from 150 mm to 1,300 mm. The northeast region is commonly known as the Drought Polygon. The rainy season generally lasts for 3 to 5 months. The temperature varies little during the year (23 to 27° C), with an average of 2,800 hours of sunshine per year. Average annual relative humidity is approximately 50 to 60%.

Vegetation

The principal range types of this region are typical of desert and woodland shrub in semiarid zones. The local name for this general range type is *caatinga*.

Forage species include *Mimosa caesalpiniaefolia, Phaseolus lathyroides, Zizyphus joazeiro, Echinochloa crusgalii, Aristida setifolia, Stylosanthes guianensis* and *S. augustifolia, Panicum parvifolium,* and *Desmodium discolor.* Others include *Spondias tuberosa, Bauhinia forficata, Aspidosperma macrocarpum, Cnidoscolus phyllacanthus, Calliandra depauperata, Ruellia asperula, Caesalpinia ferrea, Pithecolobium dumosum, Mimosa nigra, Gomphrena demissa,* and *Ipomoea glabra.*

PRODUCTION SYSTEMS USED IN NORTHEAST BRAZIL

Most livestock are allowed to graze freely in the unfenced *caatinga.* Supplementation is uncommon, even during extended dry seasons when little grazing is available. A survey of the state of Bahia indicated that fewer than 10% of the producers provided any supplements (Bahia 1975).

This same survey revealed that fewer than 8% of the producers kept only sheep and/or goats; 48% combined sheep and goats with crops, such as corn, beans, and sisal; 11% kept sheep, goats, and cattle; and 31% had sheep, goats, cattle, and crops. Of the survey population, 61% cited sheep and goats as their principal economic activity; 32%, cropping; and 5%, cattle raising. Grade breeding stock having no pedigree or performance information was used by 83% of the surveyed producers. Of the recognized breeds used, the most common was Bergamasca.

Producers generally did not have access to scientific information on nutrition, management, herd health, or genetic improvement. In addition to lack of technical information, other constraints cited were: 25% of the producers had difficulty obtaining credit, 20% did not have access to improved sires, and 7% lacked ownership of land. However, lack of technical assistance was the major constraint cited by 41% of the producers. Producers do not have ready access to vaccines and anthelmintics. Facilities are inadequate for proper control and protection of stock and hygiene is poor. As a consequence, mortality rates are high.

Some farmers actively seek technical and economic assistance, as there has been some government assistance in development of plans for facilities to improve management of sheep and goats.

The principal reason for raising sheep and goats is to provide meat, primarily for local use by low-income, rural populations. A secondary objective is to produce skins for sale to commercial tanneries. Generally, flock numbers are small, with 20 to 50 head per farm; however, some flocks have 400 to 500 head. Kasprzykowski and Nobre (1974) stated that Brazilian goat and hair sheep populations are concentrated in the Drought Polygon area with a density of 11 cattle, 8.3 sheep, and 11.4 goats per km².

Table 2.7.2 lists the approximate age and sex composition of flocks in the Northeast.

TABLE 2.7.2.　　AGE AND SEX COMPOSITION OF SHEEP FLOCKS IN THE NORTHEAST

Category	Percent
Rams	2.5
Breeding ewes	41.8
Lambs, less than 12 mo	
Males	16.6
Females	16.1
Yearlings, greater than 12 mo	
Males	9.6
Females	13.4

Source:　Sinopse Estatística do Brasil (1977).

Most animals graze exclusively on native pasture that is characterized by trees, shrubs, and *extrato herbaceo* vegetation. Irregular rainfall limits the carrying capacity of this pasture. Grasses and legumes in the northeastern pastures have high nutritional value during the rainy season, but their value declines rapidly in the dry season, causing substantial losses in livestock performance in the region.

Preservation and storage of improved pasture are not common practices. Some pasture is irrigated using water from small artificial lakes; however, the amount grown is not sufficient to supplement the limited grazing of the entire flock during the critical season. Grazing may be supplemented during the critical season with grain, cut grass, cereal bran, crop residues, silage, hay, cottonseed, cactus, sisal bulbs, legumes, or umbuzeiro leaves.

Producers rarely feed concentrates to penned animals; this practice is limited to a few farms that specialize in producing breeding stock. Mating is generally allowed throughout the year; however, parturitions tend to occur more frequently in June or July and January or February.

Ewes generally lamb in the open pasture. Losses of lambs due to navel infection and predation are high. Separation of sexes is not commonly practiced; thus, there is little selection of superior males, and ewes may lamb at young ages and light weights.

Health problems include chronic gastrointestinal parasitism (*Haemonchus, Trichostrongylus, Oesophagostomum*), coccidiosis, pediculosis, foot rot, caseous lymphadenitis, tetanus, contagious ecthyma (sore-mouth), and foot-and-mouth disease.

According to Kasprzykowski and Nobre (1974), no detailed study has been made of sheep marketing. However, they reported that the marketing process does not have many middlemen because of the low value of sheep per head.

Sheep are marketed in small groups. Prices are on a per-head basis and

TABLE 2.7.3.   SHEEP PRODUCTIVITY IN NORTHEAST BAHIA

Average number lambs, first parturition	1.0
Average number of lambs at the following parturitions	1.1
Mortality rate, %	42.0
Weaning rate, %	62.0
Average slaughter age, months	15.0
Average slaughter liveweight, kg	20.0

Source: Bahia (1975).

based on an average weight, but scales are not used. Marketed animals usually are more than a year old, with carcass weights of 10 to 12 kg. Relatively more sheep meat is consumed in small towns than in the big cities. Occasionally, the price paid the producer for the live animal equals the carcass value. The buyer's profit is the hide value, which is about 30% of the total value of the animal. Kasprzykowski and Nobre (1974) found that sheep and goat meat sells at 50 to 67% of the price of beef.

In the major cities, there is a middleman between the producer and the slaughterer whose function is basically transportation. He buys from the farms and moves the animals to the slaughter center.

For the skin market, there is a well-defined market structure. The tannery or exporter generally is located in a big city, with a network of buyers around the region. In some places, the middleman buys the skin from the slaughterer and resells it to the tannery or exporter. Export prices vary with international demand.

Skins may be graded individually or sold in large lots. The graded system has skins of two classifications: first-quality skins must weigh approximately 450 to 900 g and must have no imperfections; skins having some imperfections are second-quality and sell at one-half the price of first-quality skins. Skin imperfections include scratches, wire cuts, pock marks, holes, and decay from remnants of meat or fat. In the large-lot system, all skins are given the same average values.

PRODUCTION CHARACTERISTICS

Low flock productivity has been reported for sheep grazing extensively in the northeast region of Brazil. Table 2.7.3 was based on results from a survey of Bahia State producers (Bahia 1975); these data indicate that productivity is low and mortality is high with the traditional production systems.

In recent years major governmental participation in research on systems of tropical sheep production has increased the productivity of Brazilian hair

TABLE 2.7.4.   SANTA INÊS AND MORADA NOVA REPRODUCTIVE TRAITS (1978)

Trait	Morada Nova	Santa Inês
Number	23	21
Ewes conceiving/ewes exposed, %	95.6	80.9
Ewes lambing/ewes exposed, %	91.3	76.2
Ewes aborting/ewes exposed, %	4.3	4.8
Lambs born/ewes lambing, %	176	125
Litter size		
Single	6	12
Twins	14	4
Triplets	1	–
Gestation length, days ($\bar{x} \pm$ SE)	149.4 ± .4	150.6 ± .6

Source: Centro Nacional de Pesquisa de Caprinos, EMBRAPA (1979b).

sheep. Research centers such as the Centro Nacional de Pesquisa de Caprinos of EMBRAPA actively support research on Brazilian hair sheep.

REPRODUCTION CHARACTERISTICS

Research to evaluate two improved hair sheep races (Santa Inês and Morada Nova) provided the results shown in Table 2.7.4

In this experiment, ewes from each breed grazed (together with goats) on native pasture. Grazing density was 1.6 ha per 30 kg animal. The animals received a mixture of bone meal and common salt ad libitum. They were wormed with levamisol when inspection of manure revealed 800 eggs per gram. The lambing season coincided with the food-scarcity season (July to September 1978). There was an advantage for the Morada Nova over Santa Inês for prolificacy.

In addition to the results in Table 2.7.4, EMBRAPA (1979b) has studied reproductive traits of Morada Nova, Santa Inês, and Brazilian Somali breeds (Tables 2.7.5 and 2.7.6). These animals are kept on an improved native grass pasture with abundant water from a small lake. On the border of this lake there is enough *"capim de planta"* (*Panicum barbinode*) to support the flock during the dry season. The animals graze freely 7.3 hours per day; the rest of the time they are confined to the corral, with access to a mixture of salt and bone meal. All newborn lambs receive prophylactic treatment, including disinfection of the navel.

All experimental sheep had been on improved temporary pasture before the experiment started. These better conditions probably improved their chances of becoming pregnant.

Reproductive traits of the Brazilian Somali have been studied. Partial

TABLE 2.7.5.   REPRODUCTIVE TRAITS FOR MORADA NOVA,
                BRAZILIAN SOMALI, AND SANTA INÊS BREEDS

	Morada Nova		Brazilian Somali		Santa Inês	
	$\bar{x}$	SD	$\bar{x}$	SD	$\bar{x}$	SD
Age at first estrus, days	214.5	38.0 (6)[a]	283.9	53.4(18)[b]	219.7	64.9(6)
Weight at first estrus, kg	20.6	0.8 (6)[a]	19.7	2.1(18)	28.6	3.3(6)
Age at first mating, days	349.9	68.0 (6)[a]	342.0	32.6 (3)[b]	286.8	59.7(6)
Weight at first mating, kg	26.7	2.8 (6)[a]	23.6	0.5 (3)	32.5	3.4(6)
Estrous cycle duration, days	16.1	1.1(48)[c]	17.5	5.1(66)	19.6	7.5(17)
Estrus duration, hours	30.4	13.9(54)[c]	27.4	7.1(81)	25.8	7.1(23)
Age at first parturition, days	497.8	67.0 (6)[a]	490.3	31.1 (3)	450.4	56.1(5)
Weight at first parturition, kg	28.7	2.7 (6)[a]	28.4	2.2 (3)	39.3	3.3(5)

[a]   Number of ewes in parentheses.
[b]   Different animals in samples observed for Brazilian Somali.
[c]   Number of cycles observed.
Source: EMBRAPA (1979b).

Table 2.7.6.   ADDITIONAL REPRODUCTIVE TRAITS FOR MORADA
                NOVA, BRAZILIAN SOMALI, AND SANTA INÊS BREEDS

	Morada Nova		Brazilian Somali		Santa Inês	
Trait	$\bar{x}$	SD	$\bar{x}$	SD	$\bar{x}$	SD
Gestation length, days	149	0	150	1.4	148	1.5
Litter size, no.						
Single	2		5		1	
Twins	4				2	
Birth wt, kg						
Males (no.)	2.5(4)	.2	3.8(2)	.2	1.9(4)	.1
Females (no.)	2.0(6)	.3	3.8(2)	.5	1.6(1)	

TABLE 2.7.7.   REPRODUCTIVE TRAITS FOR THE BRAZILIAN
                 SOMALI AT SOBRAL, CEARA, 1978

Trait	$\bar{x}$	SD
No. ewes exposed	19	–
Services/ewe	1.2	–
Gestation length, days		
Single birth (13)	149.2	1.2
Twin birth (4)	147.7	.9
Ewes lambing/ewes exposed, %	89.5	–
Abortions/ewes exposed, %	10.5	–
Litter size	1.23	
Singles, no. lambs	13	
Twins, no. lambs	8	

Source: EMBRAPA (1979b).

results of work by EMBRAPA (1979b) are summarized in Table 2.7.7 for the mating season from March 23 to May 21, 1978. Ewes grazed native pasture supplemented with salt and bone meal. Rams were greased to mark mated ewes.

GROWTH CHARACTERISTICS

Research data are scarce for growth of Brazilian hair sheep. The Associação Brasileira de Criadores de Ovinos (ABCO) began a pedigree registration service but required no information on growth characteristics.

Among the Brazilian breeds, the Santa Inês shows a better development because of its origins from the Bergamasca breed. An intact 6-month-old male of this breed can weigh approximately 40 kg, if well fed in confinement (EMBRAPA 1979b). ABCO (1977) reported that weights of adult Santa Inês males averaged 80 kg; females averaged 60 kg—some adult males in this breed weighed more than 100 kg liveweight.

Figueiredo (1978) studied the biometric characteristics of 590 Morada Nova sheep grown under extensive conditions and stratified by sex and age (Table 2.7.8). Average female weights were 31 kg—males, 38 kg. Length of rump is measured between the points of the hip and the ischium; the point of reference for width of the rump was the acetabulum at the femoral articulation.

ABCO (1977) reported that Brazilian Somali males weighed from 40 to 60 kg; females, from 30 to 60 kg. EMBRAPA (1979b) reported an average weight of 26.2 kg at parturition for a sample of 17 adult females.

The Rabo Largo has not been studied formally. ABCO (1977) cited average weights of 45 kg for males and 30 kg for females.

The Universidade Federal do Ceará (1978) reported weights for Morada

134

TABLE 2.7.8. MEANS AND STANDARD DEVIATIONS FOR WEIGHT AND LINEAR MEASUREMENTS FOR MALE AND FEMALE MORADA NOVA SHEEP

Age, teeth	Weight kg		Withers height cm		Thoracic Depth cm		Thoracic Circumference cm		Rump Width cm		Rump Length cm	
	Mean	SD	Mean	SD	Mean	SD	Mean	SD	Mean	SD	Mean	SD
Suckling												
Male (26)[a]	12.6[b]	3.1[c]	50.7	3.6	19.8	1.6	54.4	5.4	11.8	1.1	14.8	1.2
Female (36)	11.5	2.9	49.6	3.9	19.1	1.8	53.3	5.1	11.6	1.5	14.2	1.7
Milk teeth												
Male (15)	24.4	3.7	61.1	2.7	25.8	1.1	68.9	3.0	14.5	0.9	18.9	0.9
Female (43)	23.9	3.4	60.0	6.2	25.4	1.1	68.9	3.6	15.3	0.9	18.4	1.0
2 teeth												
Male (9)	25.8	3.4	61.2	3.8	26.7	1.2	71.7	2.6	15.2	0.8	19.5	0.7
Female (19)	26.6	4.1	60.1	3.2	26.3	1.4	71.0	4.4	15.7	1.2	18.7	1.0
4 teeth												
Male (4)	29.2	2.2	64.7	2.1	27.6	0.8	75.3	0.9	15.5	0.7	19.8	0.6
Female (20)	26.8	3.3	61.3	3.4	26.1	1.0	70.9	3.5	15.7	0.7	18.5	1.1
6 teeth												
Male (5)	37.4	3.6	69.2	3.1	30.1	1.3	82.0	3.4	16.4	0.8	21.7	1.0
Female (37)	29.7	3.6	61.9	3.4	27.5	1.0	74.5	3.1	16.2	0.8	19.0	0.9
8 teeth												
Male (17)	38.8	5.0	66.8	4.0	31.2	1.4	83.4	4.1	17.1	0.9	22.0	1.0
Female (74)	31.3	4.2	62.0	3.6	28.1	1.2	75.4	4.4	16.3	1.1	19.2	1.0

a Numbers of animals measured. b Mean. c Standard deviation.

TABLE 2.7.9.　　AVERAGE WEIGHTS FOR PROGENY OF FIVE MORADA NOVA RAMS, PENTECOSTES STATION

Sire	Lamb numbers	Progeny average weight, kg			
		At birth	100 days	240 days	360 days
201	162	3.0	15.9	22.6	25.6
601	113	3.0	15.6	22.6	24.5
1V	103	3.0	13.7	19.2	20.7
1A	55	2.8	15.3	22.9	24.4
10A	45	2.8	13.1	-	-

Source: Universidade Federal do Ceará (1978).

TABLE 2.7.10.　　WEIGHTS OF LAMBS BORN DURING TWO SEASONS, PENTECOSTES STATION

Lambing season	Lamb numbers	Progeny average weight, kg			
		At birth	100 days	240 days	360 days
June–August (dry)	241	2.94	13.2	21.5	23.4
Dec–Feb (rainy)	237	2.97	15.9	22.7	24.4

Source: Universidade Federal do Ceará (1978).

Nova sheep, white variety; Table 2.7.9 shows the average weights for progeny of five rams. These data include progeny born during five years (1973–1978), combined across year, sex, and type of birth. All lambs were managed similarly in each year.

There were two lambing seasons per year: at the beginning of the dry season, and at the beginning of the rainy season. Table 2.7.10 shows the weight development of animals born in each season. Lambs born during the dry season consistently were lighter than those born during the rainy season, but the difference tended to decrease from 100 to 360 days of age.

Teixeira (1977) reported results obtained with the Morada Nova in 1973 under different treatments: (A) native pasture only, (B) native pasture plus worming, mineral supplements, and 250 g of concentrate supplementation (14% crude protein) 30 days before and after birth (Table 2.7.11).

Results from Centro Nacional de Pesquisa de Caprinos compare performance of lambs from 25 Morada Nova ewes and 25 Santa Inês ewes under the same management conditions (Table 2.7.12). These animals were born at the end of the rainy season and weaned during the dry season when feed was in short supply.

In spite of the large size of mature Santa Inês sheep, weight differences

TABLE 2.7.11.　COMPARATIVE PERFORMANCE OF MORADA NOVA LAMBS BORN TO EWES UNDER TWO TREATMENTS

Traits	Treatments	
	A	B
Birth weight, kg	2.9	3.2
Weaning weight, kg	9.5	16.5
Average weight at 224 days, kg	19.0	26.0
Days to reach 16 kg	168.0	84.0

Source:　Teixeira (1977); numbers not reported.

TABLE 2.7.12.　MEANS AND STANDARD DEVIATIONS FOR WEIGHTS FOR MORADA NOVA AND SANTA INÊS LAMBS, kg

Breed	Trait	Birth			210 days		
		No.	$\bar{x}$	SD	No.	$\bar{x}$	SD
Morada Nova	Female						
	Single	3	2.6	.4	3	13.1	1.1
	Twin	17	2.3	.3	9	10.5	1.4
	Male						
	Single	3	3.4	.6	3	14.0	1.4
	Twin	11	2.3	.3	7	9.8	1.3
Santa Inês	Female						
	Single	6	3.0	.7	4	11.7	3.8
	Twin	6	2.5	.2	2	13.4	2.2
	Male						
	Single	6	3.6	.3	5	15.0	3.4

Source:　EMBRAPA (1979b).

TABLE 2.7.13.　WEIGHTS OF BRAZILIAN SOMALI

Trait		Female		Male	
		Single	Twin	Single	Twin
Birth	Number	9	6	4	2
	Weight, kg	2.4	1.7	2.3	1.8
95-days	Number	9	6	4	2
	Weight, kg	12.7	8.2	12.9	9.2

Source:　EMBRAPA (1979b).

TABLE 2.7.14.   WEIGHTS OF MORADA NOVA SHEEP, QUIXADA
STATION, CEARA (1976-78)

Trait		PN	PNM	PA	PNMSE	PNMCE
				Types of pasture[a]		
Birth	Number	80	142	290	115	168
	Weight	2.0	2.0	2.0	2.0	2.1
Weaning	Number	44	86	151	105	110
	Weight	13.9	14.6	13.3	13.5	12.7

[a] PN = Native pasture.
PNM = Native improved pasture.
PA = Introduced pasture.
PNMSE = Introduced pasture without exclusions.
PNMCE = Introduced pasture with exclusions.

Source: EPACE (1976, 1977, and 1978).

between lambs of the two breeds were small and inconsistent; this could have been due to the food scarcity during the dry season.

Brazilian Somali are smaller sheep than Morada Nova and Santa Inês. In the experiment that provided data for Table 2.7.13, Brazilian Somali sheep were maintained on native grass fertilized with manure; a small lake served to preserve green material during the dry season. These animals were born during the dry season and were weaned at 95 days of age.

Additional information on Morada Nova weights and other characteristics were available from research by EPACE (Empresa de Pesquisa Agropecuária do Ceará). These are results from three years of pasture research (1976-1978). The research had 5 pasture treatments. Mating was between December 15, 1977, and February 15, 1978; lambs were weighed at 114 days of age. Table 2.7.14 shows some results of this research. Pasture type had little effect on 114-day weight; however, this experiment did not account for differences in stocking rate.

CARCASS CHARACTERISTICS

Little is known about the carcass characteristics of hair sheep in the northeast. To develop some baseline information, sheep were purchased on the free market and slaughtered at a local slaughterhouse. Table 2.7.15 lists the results of carcass evaluation.

TABLE 2.7.15.   CARCASS CHARACTERISTICS

	Santa Inês[a]		Crioula[a]		Crioula[b]		Crioula[c]	
	$\bar{x}$	SD	$\bar{x}$	SD	$\bar{x}$	SD	$\bar{x}$	SD
Number	8		8		7		10	
Live weight, slaughter, kg	44.7	2.9	26.8	2.4	24.7	3.4	16.3	3.5
Skin weight, kg	4.0	0.4	2.5	0.3	2.7	0.3	2.6	0.8
Hot carcass yields, %	46.2	3.9	39.1	3.2	41.2	1.9	43.0	1.9
Meat/bone ratio	3.0	0.5	2.6	0.5	2.5	0.5	1.8	0.2
Carcass length, cm	69.4	2.1	61.4	1.5	61.0	2.5	51.7	4.0
Leg length, cm	41.1	0.9	36.0	1.6	35.9	1.3	34.8	2.3
Thoracic depth, cm	22.8	1.0	18.5	0.9	18.3	1.2	15.9	1.4
Thigh circumference, cm	38.9	2.5	30.8	1.9	30.2	2.3	26.2	2.2
Thigh thickness, cm	10.4	0.4	7.7	0.8	7.6	0.5	5.9	0.7
Half carcass pistol cut,[d] kg	4.8	0.6	2.4	0.5	2.3	0.4	1.6	0.4
Bone, kg	2.6	0.3	1.3	0.3	1.4	0.2	1.1	0.3
Meat, kg	7.6	1.1	3.5	0.7	3.3	0.8	2.0	0.5

[a] Old ewes (full mouth, worn incisors).
[b] Old pregnant ewes (full mouth, worn incisors).
[c] Milk tooth, ram lambs.
[d] Leg and loin up to the 7th vertebra.

Source: Centro Nacional de Pesquisa de Caprinos Sobral-Ceará.

REFERENCES

Andrade, G.O.D. 1977. Alguns aspectos do quadro natural do nordeste. 75 pp. Recife: Sudene.

Anuário estatístico do Brasil. 1977. 847 pp. Fundação Instituto Brasileiro de Geografia e Estatística (IBGE). Rio de Janeiro.

Associação Brasileira de Criadores de Ovinos (ABCO). 1977. Regulamento do Registro Genealógico provisória de ovinos no Brasil. Flock-book brasileiro (F.B.B.). 30 pp. Bagé-R.S.

Associação Brasileira de Criadores de Ovinos (ABCO). 1978. *Revista ovinocultura.* 10:(22) Bagé-R.S.

Bahia. Secretaria da Agricultura. 1975. S.E.R. *Aspectos da produção e da commercialização de caprinos e ovinos no região Nordeste da Bahia.* 104 pp. Salvador.

Domingues, O. 1954. Sobre a origem do carneiro deslanado do Nordeste. Publicação no. 3 da seção de Fermento Agricola do Ceará. 28 pp. Fortaleza-Ce.

Empresa Brasileira de Pesquisa Agropecuária (EMBRAPA). 1974. Anteprojeto para Implantação do Centro Nacional de Pesquisa de Caprinos e Ovinos. 61 pp. Brasília.

EMBRAPA Centro Nacional de Pesquisa de Caprinos. 1978. Relatório Técnico de acompanhamento dos subprojetos "Estudo do efeito da vermifugação mineralização e suplementação alimentar sobre a produtividade de ovinos, e sobre a produtividade de caprinos." (Relatório interno mimeografado.) Sobral-Ce.: Fazenda Periperi S/A.

EMBRAPA Centro Nacional de Pesquisa de Caprinos. 1979a. Relatório Técnico de acompanhamento a Dr. Mason am visita aos trabalhos de pesquisa em andamento nas estações esperimentais do Nordeste. (Relatório interno grafada.) Sobral-Ce.

EMBRAPA. 1979b. Relatório Técnico Anual do Centro Nacional de Pesquisa de Caprinos e Ovinos tropicais. 54 pp. Sobral-Ce.

Empresa de Pesquisa Agropecuária do Ceará (EPACE). 1976. Relatório das atividades do projeto Caprinos e Ovinos. 46 pp. Fortaleza-Ce.

EPACE. 1977. 37 pp.

EPACE. 1978. 40 pp.

Figueiredo, E.A.P. 1978. Descrição da população de animais de raça Morada Nova. Sobral, Empresa Brasileira de Pesquisa Agropecuária/Centro Nacional de Pesquisa de Caprinos. 12 pp. (Mimeografado.) Sobral-Ce.

Kasprzykowski, J.W.A. and J.M.E. Nobre. 1974. *Possibilidades da caprinocultura a ovinocultura do Nordeste.* 182 pp. Fortaleza-Ce.: BNB/ETENE.

Mason, I. L. 1980. *Strengthening agricultural research in Brazil.* Final Report presented to the Interamerican Institute of Agricultural Sciences, Sobral, Empresa Brasileira de Pesquisa Agropecuária. 30 pp. (Mimeografado.) Centro Nacional de Pesquisa de Caprinos.

Mendonça, A. S. 1951. O carneiro rabo largo e sua introdução na Bahia, Salvador, SAIC. SAIC Boletin 4. 6 pp. Bahia.

Sinopse estatística do Brasil. 1977. 626 pp. Fundação Instituto Brasileiro de Geografia e Estatística (IBGE). Rio de Janeiro.

Teixeira, F.J.L. 1977. A herança: O meio ambiente e o melhoramento dos ovinos do Nordeste. Palestra proferida em 20/12/77 no. BNB. 14 pp. Fortaleza-Ce.

Universidade Federal do Ceará. 1978. *Relatório Técnico Retrospective das atividades do conveñio BNB/FCPC—Programa de Pesquisa em forragicultura, melhoramento e nutrição animal.* 110 pp. Fortaleza-Ce.

SHEEP PRODUCTION IN TOBAGO WITH SPECIAL REFERENCE TO BLENHEIM SHEEP STATION

Raj K. Rastogi and K.A.E. Archibald,
The University of the West Indies, Trinidad
M. J. Keens-Dumas, *Ministry of Agriculture,*
Lands and Fisheries, Trinidad and Tobago

Sheep have been raised by small farmers in the Commonwealth Caribbean countries since the days of colonization. Today, Guyana and Barbados have relatively large sheep populations and the Republic of Trinidad and Tobago has 7,000 head of sheep, of which about 5,000 are in Tobago.

This region's great demand for mutton is met by continuous importation from outside the Caribbean. In Trinidad and Tobago, local sheep production satisfies only about 15 percent of the home market for mutton (see Central Statistical Report, 1977). Recently, however, Trinidad and Tobago (and other Commonwealth Caribbean countries) have assigned higher priorities to producing meat from small ruminants. Tobago shows excellent promise for exploitation of sheep to increase local supplies of mutton.

This report critically examines the status of sheep production in Tobago in general, and at the Blenheim Sheep Station in particular. The information was gathered from library sources, personal interviews, visits to private sheep farms, and through analysis of data collected at the Blenheim Sheep Station.

GEOGRAPHICAL BACKGROUND

Trinidad and Tobago are two islands in the southernmost part of the Caribbean Sea. Tobago, the smaller of the two islands, has an area of about 301 km² and lies 32 km to the northeast of Trinidad at 11° 15′ N and 60° 40′ W. Geologically, Tobago is an extension of the Venezuelan or Caribbean coast range and the northern range of Trinidad. The human population of Tobago is about 45,000.

Tobago's climate is humid-tropical, with relative humidity varying from 80 to 85%. The climate is tempered by sea breezes that moderate the daily temperatures that usually range between 25° and 31° C. Temperatures vary most between day and night.

There is a major dry season from January to May and a short dry season in October. The mean annual rainfall is 1,700 mm.

HISTORY AND DESCRIPTION OF SHEEP IN TOBAGO

Sheep production has been practiced in Tobago for well over 100 years. There were no indigenous sheep in Tobago and the present sheep are descendants of imported breeds.

The following history of sheep on the Tobago Government Farm since 1900 is based on the Annual Reports of the Department of Agriculture, Trinidad and Tobago.

As early as 1909, a half-breed woolless ram and two woolless ewes in lamb to a purebred woolless West African ram were imported from Barbados. Around 1912, a few Romney Marsh sheep were imported from the United Kingdom. However, neither breed performed well, and by 1915 there were no sheep left at the Government Farm. In 1920, the Tobago Government Farm resumed sheep production with six ewes purchased locally from the Lowland Estate. A purebred Southdown ram was imported from England but died one year later, leaving descendants. Afterwards, two West African rams were reintroduced.

About 1930, 72 ewes and 8 rams of the Blackhead Persian breed were imported from South Africa into Trinidad for distribution throughout the West Indies, Demerara (now Guyana), and Venezuela. Early in 1931, these sheep were sold to private owners in Trinidad and Tobago. The Government Farm in Trinidad kept five ewes and two rams; one of these rams was later sent to the Tobago Government Farm. By 1936, the Blackhead Persian sheep had earned a good reputation as a meat breed, and rams of this breed were mated with local ewes to produce market lambs of good conformation. By the end of 1948, the Tobago Government Farm had a total of 55 sheep of Blackhead Persian and West African types. About 1950, yet another breed, the Wiltshire Horn, was introduced from Scotland, but it was not productive and soon disappeared.

The date of the first importation of Barbados Blackbelly sheep to Tobago is not known; however, by 1953 the breeding policy of the Tobago Government Farm was geared toward the gradual establishment of a flock of sheep graded to the Barbados Blackbelly breed. By 1958, three breeds of sheep were being maintained, viz., Blackhead Persian, "West African," and Barbados Blackbelly. Sometime in the late 1960s or early 1970s, sheep of the Suffolk breed also were introduced into Tobago. More recently, importations of Barbados Blackbelly sheep have been made from Antigua (in 1976) and Barbados (in 1977-1978).

The present sheep population on Tobago consists of the Blackhead Persian, West African, Barbados Blackbelly, and grades of sheep that resulted from indiscriminate crossing among the introduced breeds. Of the three pure breeds, there appear to be fewer Blackhead Persian. The sheep population exhibits hair coat-color patterns that vary from solid white to shades of brown to red to mottled to dark brown (or black). All sheep are of the hair type, with some of the grades showing traces of wool as a result of crossing with wool sheep. All sheep on Tobago are polled.

SYSTEMS OF PRODUCTION AND MANAGEMENT

The majority of sheep on Tobago are raised by small farmers in flocks of 1 to 5 animals as a ready source of cash or meat for special occasions. Most farmers use the tethering system whereby sheep are pegged to a rope about 3 to 5 m long to feed on roadside grazing or on communal pastures. The peg location is changed once or twice daily to give the sheep access to a fresh browsing area. Sheep graze for about 7 hours every day, starting at 0700. Family labor is used to manage the sheep. Few concentrates are provided, except for household scraps and other supplements that might be fed during an extended dry season. The sheep are provided with varying kinds of housing during the night, usually in the immediate vicinity of the farmer's home. Housing may be only a shed erected at ground level and made from corrugated iron sheets, wood, or leaves.

Apart from a few large flocks (50 to 150 ewes and followers) kept by private estates, the only sheep "station" in the country is the government farm at the Blenheim Estate. These larger flocks are raised under an extensive system whereby sheep are allowed free grazing on partially improved pastures during the day and are penned during the night for protection against predators and larceny. Widespread larceny and attacks by dogs are strong deterrents to sheep raising in Tobago and present a serious barrier to the expansion of the sheep population.

Owners of large flocks may supplement grazing with concentrates, especially during the dry season. Agricultural and industrial by-products are not produced in Tobago and must be imported from Trinidad. Thus the supply of concentrates to the Tobago market is very irregular.

Little is known about the general health status of sheep on Tobago; however, the sheep appear to be free of any major disease. Gastrointestinal parasites seem to be the major health problem. Other common diseases are pneumonia and foot rot.

Although the present level of husbandry is suboptimal, improved breeding, feeding, and management can greatly increase the potential contribution of Tobago's sheep.

MARKETING

The marketing and slaughtering of sheep is not well organized. Sheep are generally sold "on the hoof" in close proximity to the point of production—either to neighbors, to butchers for slaughter in Tobago, or to market traders (middlemen) for shipping to Trinidad for slaughter. The Tobago butchers visit the farms regularly to purchase the sheep that usually are slaughtered at the farm or, rarely, at the two municipal slaughterhouses. The majority of surplus sheep in Tobago are shipped alive to Trinidad for slaughter there. The skins are not utilized in Tobago; there is no tannery on the island.

Although the number of sheep slaughtered in Tobago and the number shipped to Trinidad are not recorded, estimates based on the sheep population in Tobago would suggest that Tobago markets about 1,500 sheep annually. The selling price of sheep is based on size and is determined by visual appraisal of the animal; price is often biased in favor of the butcher. The price of fresh mutton in the Tobago market in 1979 was about $3 (US) per kg carcass weight.

BLENHEIM SHEEP STATION

In 1971, a 40-ha sheep breeding, multiplication, and demonstration project was developed from the old Blenheim Estate on the windward coast of Tobago. The topography of the area varies from hilly to mildly undulating, with slopes of 10 to 30°.

Approximately 24 ha are in Pangola grass (*Digitaria decumbens*) pastures and 1 ha in elephant grass (*Pennisetum purpureum*); these are usually harvested for hay or silage. The Pangola pasture is subdivided into paddocks of 0.4 to 1.6 ha each, which are grazed rotationally for a period of 7 to 9 days followed by a regrowth period of 4 to 6 weeks. The pasture is fertilized at the rate of 25 kg N/ha/annum, applied in two or three applications. Pasture productivity is poor, both in terms of quantity as well as the quality of grass produced, and the station is somewhat overstocked.

Daily, throughout the year, ewes are let out to pasture at 0800 and are brought in at 1500, to remain indoors overnight. Ewes are confined only at lambing for a period of 2 to 5 days. If available, concentrates are fed to the ewes at the rate of 0.5 kg/head/day for a period of 3 to 4 weeks prior to and after lambing.

Rams are kept in individual pens at the nearby Hope Farm. They are fed freshly cut grass ad libitum, without concentrates.

Lambs are pastured with their dams as soon as possible after birth and allowed to suckle as long as the ewe will permit. Lambs are not fed concentrates, but begin to eat grass at about 4 weeks and are usually weaned by 12 weeks of age.

All sheep are dewormed routinely. A foot bath is used for dipping to prevent foot rot.

Until 1975, mating was allowed throughout the year. However, to increase lamb survival, a seasonal program of breeding was begun so that lambing occurs during the dry season from mid-January to mid-March. Normally, the breeding policy is to maintain pure breeds, except for some crossbreeding to reduce the inbreeding level. As of January 1979, the station had 448 sheep, including 365 adult breeding ewes, 12 breeding rams, and 71 lambs. The Barbados Blackbelly breed constituted 38.2% of the total; the West African, 19.6%; the Blackhead Persian, 9.6%; and various crossbred sheep, 32.6%.

SHEEP PERFORMANCE AT THE BLENHEIM STATION

Performance data at the Blenheim Station, which have been recorded irregularly, are summarized below.

Growth

Rastogi et al. (1979) analyzed body weights for lambs (from birth to weaning at 12 weeks of age) for the four breeds born at Blenheim Station during October 1974. The results shown in Table 2.8.1 indicated no significant differences due to breed effect. Poor nutrition may have limited the expression of full genetic potential for growth. Data in Table 2.8.1 are comparable to those reported by Rios (1968), Devendra (1975), Reverón and Garcia (1975), and James (1976).

Mature Weight of Ewes

All ewes were weighed in December 1978 and again in March 1979; Table 2.8.2 summarizes the results. Breed of ewe and month of weighing were not strongly related to mature ewe weight, although there was a tendency for March weights to be slighly lower, probably due to the effect of the dry season. The Blackhead Persian ewes were slightly lighter (27.6 kg) than those of the other breeds, while Barbados Blackbelly ewes were marginally heavier (32.3 kg). The data for Blackbelly ewes are similar to those reported

TABLE 2.8.1.   BREED MEANS AND STANDARD ERRORS FOR LAMB
                  WEIGHTS AND GAIN

| | Body weights, kg | | | | | | Daily gain, g | | |
| | Birth | | | 12 wks | | | Birth to 12 wks | | |
	No.	$\bar{x}$	±SE	No.	$\bar{x}$	±SE	No.	$\bar{x}$	±SE
Blackbelly	29	3.1	+ 0.1	27	13.8	+ 0.6	27	127	+ 6
Grade Blackbelly	29	2.8	+ 0.1	26	13.8	+ 0.6	26	130	+ 6
West African	31	2.9	+ 0.1	23	14.9	+ 0.7	23	141	+ 7
Blackhead Persian	9	3.0	+ 0.2	9	14.4	+ 1.1	9	136	+11

Source: Rastogi et al. (1979).

TABLE 2.8.2.   MATURE WEIGHT BY BREED OF EWE AND
                  MONTH OF WEIGHING [a]

| | Mean body weight, kg | | | | | |
| | Dec   1978 | | March   1979 | | Overall | |
	No.	$\bar{x}$	Range	$\bar{x}$	Range	$\bar{x}$	Range
Barbados Blackbelly	98	31.9	24–42	32.7	28–45	32.3	24–45
West African	64	30.8	18–45	29.2	18–41	30.0	18–45
Blackhead Persian	29	28.5	20–36	26.8	18–36	27.6	18–36
Mixed grades	49	32.2	22–45	29.3	20–38	30.7	20–45

[a]   Ewes of different breeds were of comparable average age of three
      years.

by Maule (1977) and Mason (1978). On the other hand, Devendra (1977)
reported somewhat higher body weights for adult Blackbelly and West
African ewes.

Reproductive Performance of Ewes

Table 2.8.3 summarizes the litter sizes recorded for some of the ewes from
each breed during 1976 and 1977 lambings. Litter size of the Blackhead Persian ewes was less than that of ewes of the other three types. The litter sizes
of Blackbelly and West African ewes shown in Table 2.8.3 are lower than
those found in the literature (Bodisco et al. 1973; Mazzarri et al. 1973 and
1976; Devendra 1977).

TABLE 2.8.3.   DISTRIBUTION OF LITTER SIZE FOR BLACKBELLY, WEST AFRICAN, BLACKHEAD PERSIAN, AND MIXED-BREED EWES

Breed	Year	No. ewes lambing	Litter Size % ewes bearing 1	2	3	Mean
BB	1976	50	72.0	28.0	-	1.28
	1977	18	55.6	33.3	11.1	1.56
	Total	68	67.6	29.4	3.0	1.35
WA	1976	105	74.3	25.7	-	1.26
	1977	25	72.0	28.0	-	1.28
	Total	130	73.8	26.2	-	1.26
BP	1976	24	91.7	8.3	-	1.08
	1977	14	85.7	14.3	-	1.14
	Total	38	89.5	10.5	-	1.10
Mixed	1976	75	80.0	18.7	1.3	1.21
	1977	28	67.9	32.1	-	1.32
	Total	103	76.7	22.3	1.0	1.24

In a separate study conducted at the Blenheim Station, Keens-Dumas (1977, unpublished) allowed a group of 42 adult Barbados Blackbelly ewes (imported from Antigua in 1976) to run with Blackhead Persian rams for a period of six weeks, from mid-January to mid-March 1977. Ewes were kept on pastures and were fed only small amounts of concentrates. As the ewes neared lambing, they were transferred to a new location at Kendal Farm School due to lack of grazing at Blenheim. Lambing began in early July 1977 and continued until mid-August. After lambing, ewes were fed 225 g of dairy concentrate ration per head per day until the 12th week of lactation. The following observations were made:

Ewes mated	42	Single births	68.6%
Ewes lambing	35 (76.1%)	Twin births	31.4%
Lambs born	46	Lamb mortality	
Lambs born/ewe	1.31	at 12 weeks	38%

Table 2.8.4. LAMB AND EWE MORTALITY OVER THREE YEARS

Item	Year 1976	1977	1978	Overall
Lambs:				
No. born	488	370	336	1194
No. died	150	202	114	466
% mortality	30.7	54.6	33.9	39.0
Ewes:				
No. in flock	413	538	506	1457
No. died	30	111	109	250
% mortality	7.3	20.6	21.5	17.1

Lamb Mortality

Table 2.8.4 summarizes the data collected on lamb mortality from 1976 to 1978 including preweaning and postweaning death losses. Overall lamb mortality was 39% and ewe mortality was 17.1%. Specific causes for high mortality were not identified, but poor nutrition, herd health, and herd management probably contributed to the high mortality.

SUMMARY AND CONCLUSIONS

Sheep production in Tobago is an important component of animal agriculture with excellent potential. Tobago's sheep (approximately 5,000 head) are raised for mutton production and are sold to neighbors or butchers, or they are shipped alive to Trinidad for slaughter. The skins are not utilized; there is no tannery on the island.

Tethering and extensive systems of production and management are used, and the level of husbandry is generally low.

Sheep are usually of three breeds: Barbados Blackbelly, West African, and Blackhead Persian, along with some sheep of mixed origin. Blackbelly and West African sheep are prolific and are fertile year-round. (There is little information regarding a low or high season for sexual activity.) The Blackhead Persian is better liked for its conformation. Breed differences in weight at birth and 12 weeks and in daily gain from birth to 12 weeks were not significant. The number of lambs born per ewe lambing was 1.35 for Blackbelly, 1.26 for West African, 1.24 for crossbreds, and 1.10 for Blackhead Persian ewes.

Improved sheep husbandry and breeding offer considerable promise for increasing total productivity. Improvement of pure breeds by selection will require a clear definition of biological objectives. The comparative performance of the various breeds must be analyzed under the existing systems of production, and systems of crossbreeding must be developed that will exploit the fertility and prolificacy of local hair sheep, as well as the mutton-producing ability of the exotic breeds.

ACKNOWLEDGMENTS

The authors are indebted to Mr. D. Davis of Blenheim Sheep Station for his help in collection of the data, to Mr. E. Bailey for analyzing some of the data, and to Dr. F. G. Youssef of The University of the West Indies for reading the manuscript. We are also grateful to the Ministry of Agriculture, Lands and Fisheries, Trinidad and Tobago, for permission to use their data.

REFERENCES

Bodisco, B., C. M. Duque, and A. Valle S. 1973. Comportamiento productivo de ovinos tropicales en el período 1968–1972. *Agronomía Tropical* 23(6):517–540.

Central Statistical Report. 1977. No. 4. Central Statistical Office, Government of Trinidad and Tobago.

Devendra, C. 1975. Sheep and goat production in Guyana. *Z. Tierzuchtg. Zuchtgsbiol* 92:305–317.

Devendra, C. 1977. Sheep of the West Indies. *World Review of Animal Production* 13(1):31–38.

James, L. A. 1976. Cattle and sheep production data from ECCM countries and Barbados. Mimeographed report (26 pp.) submitted to Caribbean Rural Development Advisory and Training Service, St. George, Grenada.

Mason, I. L. 1978. Conservation of animal genetic resources. Report on a visit 4 January–14 February 1978 to Barbados, St. Croix, Dominican Republic, Haiti, Jamaica, Cuba, Mexico, and U.S.A. Mimeo. FAO, Rome, Italy.

Maule, J. P. 1977. Barbados Blackbelly sheep. *World Review of Animal Production,* 24:14–23.

Mazzarri, G., C. E. Fuenmayor, and C. M. Duque. 1973. Control del ciclo estral mediante el uso de esponjas vaginales impregnadas en acetato de fluorogestona en ovejas. *Agronomía Tropical* 23(23):315–321.

Mazzarri, G., C. E. Fuenmayor, A. E. Reverón, and M. Arriojas. 1976. Sincronización y fertilidad en ovejas tropicales en diferentes épocas del ano e intervalos entre parto y servicio. Mimeo. 7 pp. Instituto Investigaciones Zootécnical, Maracay, Venezuela.

Patterson, H. C. 1976. The Barbados Blackbelly sheep. 19 pp. Bulletin of Barbados Ministry of Agriculture, Science and Technology, Bridgetown, Barbados.

Rastogi, R., F. G. Youssef, M. J. Keens-Dumas, and D. Davis. 1979. Note on early growth rates of lambs of some tropical breeds. *Tropical Agriculture* (Trinidad) 56(3):259–261.

Reverón, A. E. and I. Garcia. 1975. Comportamiento en el trópico de tres razas africanas de ovejas y sus mestizas con la oveja Criolla. Mimeo. 24 pp. Biblioteca CENIAP, MAC, Maracay, Venezuela.

Rios, C. F. 1968. Behaviour of Persian Blackhead, West African, Barbados Blackbelly and native Criollo at the Centro de Investigaciones Agronómicas, Venezuela. *Proceedings Second World Conf. Anim. Prod.* (p. 5.4), Univ. of Maryland, U.S.A.

BARBADOS BLACKBELLY AND CROSSBRED SHEEP PERFORMANCE IN AN EXPERIMENTAL FLOCK IN BARBADOS

Harold C. Patterson, *Ministry of Agriculture, Barbados*

INTRODUCTION

In Barbados, sheep traditionally have served as a source of ready cash for small households or farmers; they usually are kept on land that supports no other agricultural enterprise. Most of the sheep are raised in the drier districts of the north, east, and south of the island, especially near the coast in the parishes of St. Peter, St. Lucy, St. Andrew, St. Philip, and Christ Church.

The 1971 census of agriculture listed a population of 27,064 sheep owned by 9,728 households. Of these sheep, 22,089 (81.6%) were owned by 8,526 small producers (87.6%), each of whom owned 0.45 ha of land or less (5,841 of these producers each had less than 0.05 ha of land and were classified as "landless").

Thus, the majority of sheep in Barbados are owned by persons whose major means of securing pasturage and roughage is by grazing their sheep on open lots, in parks, on other people's pastures, or along roadsides. In these cases, some supplementary feeding is provided. Many of this group are part-time farmers whose sheep production activities are not motivated by profit; rather they regard sheep as a savings or "piggy" bank.

Blackbelly sheep are not marketed through supermarkets and hotels; most are marketed through local butchers, or the owners sell the sheep directly.

The first historical reference to Blackbelly sheep in Barbados appears to be in Ligon's (1657) history of Barbados. These sheep have been on the island for more than 300 years, and a distinctive breed has evolved. Combs (Chapter 2.13) provides a detailed account of the evolution of the Blackbelly breed, suggesting that it traces to a combination of West African hair sheep and an unknown but prolific European wool breed.

Other writers suggest that, because of the preference of butchers for Blackbelly sheep and that animal's suitability to the tropics, there could have been both natural and intended selection toward the hardier hair-type sheep. This selection process could account for the fairly uniform type of sheep known today as the Barbados Blackbelly.

GENERAL BREED CHARACTERISTICS

Barbados Blackbelly sheep are covered with hair that ranges in color from light brown to dark reddish brown. They have black bellies (under surfaces) with black points on the face and legs. The face sometimes has two whitish lines running down from the forehead to the nose.

In general appearance (at a distance), these sheep resemble small hornless Jersey cattle, or deer. They are decidedly "leggy," but have fairly deep bodies and well-sprung ribs. There is a fair width of back and loin, but a very deficient hindquarter (Patterson 1976).

The rump is quite steep from the hips to the pin bones and the tail is set very low. The legs generally are quite well set and the sheep are active and lively. There is a slight tendency to a Roman nose, especially in the ram.

Compared to more recognized mutton types, Blackbelly sheep are less "squatty" and are slower growing. Weaning weights and subsequent growth rates are usually less than those of true mutton types. However, well-fed, well-grown Blackbelly rams at maturity can weigh between 150 and 200 pounds (68 and 90 kg) and mature ewes between 90 and 130 pounds (40 and 59 kg). In Barbados, the typical farmers' sheep would have lower weights nearer the lower end of these ranges (Patterson 1976).

STATION LOCATION

In 1972, a sheep station was established at Six Cross Roads, St. Philip, in the drier southeast part of the island.

Fifty-one ewes and three rams originally held at The Home Agricultural Station (approximately 1 km away) were transferred to the new station. These animals had been sent to the Home Agricultural Station as lambs from the Central Livestock Station in St. Michael.

The flock remained at Six Cross Roads, St. Philip, until September 1976, when they were moved to Greenland, St. Andrew, thus releasing the Six Cross Roads site for industrial development.

Every second year, one or two rams were brought in from the private sec-

TABLE 2.9.1.   DAILY TEMPERATURE AND MONTHLY RAINFALL
STATISTICS

Month	Average daily temperature, °C[a]		Monthly rainfall, cm[b]	
	x̄	SD	x̄	SD
January	25	.5	41	18
February	25	.6	68	35
March	25	.6	34	14
April	26	.6	44	21
May	27	.5	32	19
June	27	.4	66	72
July	27	.3	60	31
August	27	.2	142	45
September	28	.2	117	28
October	27	.4	210	50
November	27	.5	177	101
December	26	.8	138	103

[a] Means and standard deviations for monthly averages for 7 years,
1972-78.

[b] Means and standard deviations for monthly totals for 5 years,
1972-76; average annual total was 1131 cm.

tor to avoid inbreeding. Some rams also were selected within the flock, as
well as from the flock at the Central Livestock Station.

CLIMATE

As Table 2.9.1 indicates, daily temperatures are fairly constant throughout
the year in Barbados, with the extremes ranging no more than 3° C apart.
The temperatures are lowest from December through March or April. Rainfall is highly variable between months and between years.

PASTURES AND GRAZING MANAGEMENT

The former Six Cross Roads site had 8.9 ha of grassland divided into 8 paddocks of 1.1 ha each. The pasture species consisted mainly of Antigua hay
grass (*Dicanthium caricosum*), with some Pangola grass (*Digitaria
decumbens*).

Antigua hay grass is not of high quality, but the sheep perform
reasonably well on this pasture; it is a natural grass in the St. Philip area. It
grows to a height of 15 cm in the dry season when fertilized and irrigated.
During the wet season, it attains heights of 60 cm, but at this height must be
cut for efficient use by the sheep.

In 1973, the legume Siratro was introduced to test its response under normal grazing conditions. Siratro supported the sheep for a longer period than did Pangola grass, but did not recover quickly following close grazing.

Occasionally the paddocks were plowed and replanted with Pangola grass. However, the Antigua hay grass soon encroached again and eventually replaced the Pangola grass.

Irrigation Practices

In late 1973, an irrgation system was established to improve the efficiency of pasture management. Fertilizer was also applied at the rate of 250 kg/ha every two months. Even with irrigation, the pastures dried out very rapidly due to the high winds.

Grazing Period

At Six Cross Roads, shades or shelters were erected in the paddocks so that one shelter served two paddocks. The animals were grazed for about 8 hours daily from about 0700 to 1500. Animals also had free access to fresh water in the paddocks.

When the animals were brought in each evening, they were offered hay (free choice) and dairy concentrate (18% protein mixture) fed at the rate of 0.23 to 0.45 kg per animal, depending on the age and the stage of pregnancy or lactation.

At the Greenland site (to which the animals were transferred in September 1976) there were some 97 ha of improved pastures, of which about 20 ha were devoted to sheep grazing.

The pastures consisted basically of Pangola grass, with some encroachment by other species; for example, paragrass in some low-lying, poorly drained areas and some unidentified weeds, grasses, and brush in other areas. There were also about 8 ha of elephant grass (*Pennisetum purpureum*).

At Greenland, all paddocks except one had no shade. The sheep were grazed for about 6 hours from 0700 to about 1300. When the animals were brought in, they were offered hay (free choice) and dairy concentrates at the rate of 0.23 to 0.45 kg per animal depending on the age and physiological status of the animals. Water also was available, free choice.

MANAGEMENT

Breeding Flock

No distinct breeding seasons were followed at Six Cross Roads. The ewes were bred year-round. To identify the lambs sired by a particular ram, a rest period of two months was allowed before reintroducing a ram to the flock.

At Greenland, the system was somewhat different. Two distinct breeding periods were organized each year: (1) near the end of the year (from around September through January) to include the exotic sheep used for cross-breeding; and (2) after the spring lambing and during the summer months (from April to July or August). Ewes were returned to the breeding groups if not definitely known to be pregnant, thus ensuring that they would not remain unproductive for an extended period.

For ease of handling, the ewes in the various breeding groups were not pastured during the mating period. At this time, hay was fed free choice and dairy concentrate was fed at the rate of 0.45 kg per ewe. Water was offered ad libitum. At the end of the mating period, the rams were removed. Then the ewes from the different breeding groups were combined and managed as a single group. After midpregnancy, the rate of concentrate feeding was increased to 0.56 kg per animal and better pastures were provided for these ewes. Just before lambing, the ewes were again confined in the pens. Sometimes a portable lambing stall was set up in the pen for a brief period to give some protection to weaker lambs. The main danger was that lambs might be crushed by adult animals. At lambing time, the ewes received much more care and attention, especially those with multiple births. When possible, the managers ensured that the lambs suckled the first milk (colostrum).

Lambing occurred over a 2-month period. After lambing, the ewes were kept inside with their lambs and were fed concentrates and hay. After a few weeks, 0.11 to 0.23 kg of concentrate/lamb was added to the feed daily, allowing the lambs additional feed in preparation for weaning.

Weaning

The lambs were weaned about 6 weeks after the last lambs were born. At this time, some lambs were 3 months of age or older. This practice was adopted for ease of management during lambing and at weaning. Group weaning avoided the problem of providing additional space for housing individually weaned lambs. At weaning time, the lambs were removed to other pens where they were fed hay and concentrates for a short adjustment period. The lambs then were (1) selected as flock replacements for the next generation, (2) moved to the Animal Nutrition Unit for feeding trials, or (3) sold.

A lapse of 1.5 to 2 months was allowed following weaning before new matings. At weaning the lambs were weighed and dewormed; regular monthly deworming was continued.

Health

Because of the danger of reinfection by ingestion of eggs of internal parasites, the practice of bedding down the lambs was discontinued.

The most serious internal parasites encountered were *Haemonchus contortus*, nodular worms, and tapeworms. Anthelmintics were rotated regularly so that the parasites would not develop resistance to these products. Pasture rotation (which was practiced on only a limited scale because of inadequate subdivision of paddocks) is planned for use as the basis for effective parasite control in the future, as more pastures are subdivided.

The animals were treated periodically for coccidiosis, which is not a serious problem.

Other problems that occasionally occurred at the Greenland site were (1) foot rot, which particularly affected the exotic temperate sheep used in the crossbreeding experiments; and (2) pneumonia, which was a special problem when the sheep were confined during the breeding season. One of the factors contributing to foot rot and pneumonia was the dampness in the pens caused by inadequate run-off of urine and the general dampness of the area. Heat stress was another problem, especially with the exotic sheep. Shearing at frequent intervals was helpful in preventing stress.

CROSSBREEDING PROGRAMS

In 1973, some 720,829 kg of mutton and lamb meat were imported into Barbados. Local production accounts for no more than 6 to 10% of consumption. It is estimated that a breeding flock of 26,431 sheep on 4,117 ha would be adequate to supply the sheep meat needs of Barbados.

In the mid-1970s, a decision was made to import temperate breeds to improve mutton conformation of the local Blackbelly sheep. Thus, 52 sheep of two breeds, Suffolk and Dorset, were imported from the United Kingdom in October 1976.

One month after their arrival in Barbados, the exotic sheep were divided into breeding groups: three pure breeding groups (Barbados Blackbelly, Suffolk, and Dorset) and two crossbreeding groups, involving the mating of Suffolk and Dorset rams to Barbados Blackbelly ewes.

In the early stages of their acclimatization to a tropical environment, the exotic sheep were difficult to shear. The electric shears overheated constantly and manual shearing proved too tedious. The woolly animals were obviously under stress from the heat and would often pant vigorously. Fortunately, they arrived during the cooler months of the year.

Apart from heat stress, the exotic sheep were very susceptible to foot rot and occasional cases of wool rot. Although both problems were treated promptly, foot rot was a recurring problem.

During the first lambing season, only 42 lambs were produced from matings with the 40 exotic ewes, most of which were kept for multiplication purposes. Four crossbred castrates were slaughtered and test-marketed.

TABLE 2.9.2.   BIRTH WEIGHT OF BARBADOS BLACKBELLY LAMBS
                  BY SEX, LITTER SIZE, AND AGE OF DAM, kg

Litter size	Males			Females			Ratios to avg wt for single births
	No.	$\bar{x}$	SD	No.	$\bar{x}$	SD	
1	83	3.27	.07	90	3.18	.07	
2	283	3.40	.25	272	2.77	.11	.96
3	191	2.40	.04	189	2.31	.03	.73
4	29	2.13	.08	20	2.09	.09	.65

Age of dam, mo	No.	$\bar{x}$	SD
< 19	43	3.45	.68
19–35	37	3.90	.82
> 35	280	3.13	.23

Since then, 2 additional sets of animals have been slaughtered and their carcasses evaluated.

The animals studied were approximately 15 months old, which is somewhat past the lamb stage. Animals were slaughtered and hung for 7 to 10
days in a chill room at the Barbados Marketing Corporation's slaughterhouse facilities. The carcasses then were cut up, weighed, packaged, and
sold; the emphasis was on normal commercial cuts.

Objective comparison between the breeds is difficult because the Blackbelly animals were intact rams, while the crossbred exotics were castrated to
reduce the likelihood of these rams being used for breeding.

RESULTS AND DISCUSSION

Barbados Blackbelly

Table 2.9.2 shows the birth weights by sex, type of birth, and age of dam.
Twin males were heavier than single males, an unexpected result, and were
much heavier than twin females. Otherwise, the results conformed to a
general pattern of small differences between sexes and of decreasing weights
with increasing litter size. Age-of-dam effects were curvilinear.

Mean birth and weaning weights for male and female lambs were averaged for each litter size; average weights for litters of 2, 3, and 4 lambs are
shown as ratios to the average for single births (Tables 2.9.2 and 2.9.3).

Females were consistently slightly lighter than males at weaning (Table·
2.9.3) and weights of both sexes decreased substantially with increased litter
size. The effect of age of dam was again curvilinear; numbers of animals in

TABLE 2.9.3.   WEANING WEIGHT OF BARBADOS BLACKBELLY LAMBS
                BY SEX, LITTER SIZE, AND AGE OF DAM, kg

Litter size	Males			Females			Ratios to avg wt for single births
	No.	$\bar{x}$	SD	No.	$\bar{x}$	SD	
1	34	12.0	.6	45	11.6	.40	
2	109	9.7	.2	105	9.3	.24	.81
3	82	8.8	.3	62	8.1	.28	.72
4	6	8.1	.6	4	7.1	.29	.64

Age of dam, mo	No.	$\bar{x}$	SD
< 19	17	10.8	.7
19–35	3	11.8	1.3
> 35	112	9.9	.3

TABLE 2.9.4.   EWE WEIGHT, AGE AT FIRST LAMBING, LAMBING
                INTERVAL, AND NO. OF LITTERS PER EWE FOR
                BLACKBELLY EWES

Trait	Age	No.	$\bar{x}$
Ewe weight, kg	< 19 mo.	17	37.8
	19–35 mo.	34	43.1
	> 35 mo.	121	44.9
Age at first lambing, days		37	581.0
Lambing interval, days		414	256.8

No. of litters per dam	1	2	3	4	5	6	7	8	Total
No. of ewes	554	418	224	128	62	22	10	2	1420
% of total	39.0	29.4	15.8	9.0	4.4	1.5	0.7	0.1	100

two of the age-of-dam groups were quite small. The average weaning
weight overall of 9.5 kg was relatively low and was probably a factor in the
higher-than-expected postweaning mortality.

Mean ewe weight recorded at Greenland in 1979 increased from 38 kg for
18-month-old ewes to 45 kg for ewes 36 months and older (Table 2.9.4).
Two ewes in the oldest group weighed 62 kg, the highest weight recorded.

The average ewe first lambed at about 19 months of age and had a lamb-

TABLE 2.9.5.   DISTRIBUTION OF LAMBINGS AND LITTER SIZE BY
                 MONTH IN 3 FLOCKS OF BARBADOS BLACKBELLY
                 SHEEP, 1973-79 [a]

Month	No. lambings	% of total	Avg litter size
January	47	13.0	2.00
February	27	7.5	1.74
March	48	13.3	2.27
April	10	2.8	2.10
May	12	3.3	2.08
June	8	2.2	2.75
July	34	9.4	2.24
August	37	10.2	2.54
September	9	2.5	2.56
October	45	12.4	1.80
November	73	20.2	1.90
December	12	3.3	1.25
Total	362		2.10

[a] Data compiled by W. Combs, M. Hunte, and H. Patterson.

ing interval of 257 days. This lambing interval is related to the management system in which ewes were not returned to the ram until sometime after their lambs were weaned; the interval indicates that there is little delay in rebreeding when the ewes are returned to the rams. Lambing data collected in 3 flocks in Barbados from 1973 to 1979 (Table 2.9.5) support the conclusion that Blackbelly ewes reached estrus and conceived at any time of the year. Lambings occurred in every month, with no clearly identifiable seasonal pattern in incidence of parturition or mean litter size. The lambing interval of 257 days would permit 3 complete lambing cycles in about 2 years.

At the time these data were summarized, up to 8 litters had been produced by Blackbelly ewes in the Six Cross Roads, and Greenland flocks managed up to 8 litters (Table 2.9.4). However, only 6.7% of the ewes had produced more than 4 litters.

Litter size ranged from 1 to 5, with 24% of all litters having 3 or more lambs (Table 2.9.6).

Lamb survival overall was 78%, with survival of both twins and singles a little over 80% and that of triplets and quadruplets less (Table 2.9.7). Female lambs had a higher survival rate than males in litters of 2, 3, or 4 and overall, but not when born as singles. The data on litter size weaned showed that 11.8% of ewes that lambed did not raise any lambs, and 8.4% of ewes raised at least three lambs. Mean litter size weaned was 1.5 for all ewes that lambed.

TABLE 2.9.6.   DISTRIBUTION OF LITTER SIZE FOR BLACKBELLY
                   EWES, SIX CROSS ROADS STATION, 1972-76

Litter size	1	2	3	4	5	Total	$\bar{x}$	SD
No. of litters	293	490	219	28	1	1031	1.99	.78
% of total	28.4	47.5	21.2	2.7	0.1	100		

TABLE 2.9.7.   LAMB SURVIVAL BY TYPE OF BIRTH AND LITTER
                   SIZE WEANED

Litter size at birth	Males			Females			Total		
	Born	Weaned	%	Born	Weaned	%	Born	Weaned	%
1	44	38	86.4	44	34	77.3	88	72	81.8
2	151	123	81.5	143	123	86.0	294	246	83.7
3	98	64	65.3	102	73	71.6	200	137	68.5
4	18	13	72.2	12	10	83.3	30	23	76.7
Total	311	238	76.5	301	240	79.7	612	478	78.1

	Litter size weaned					
	0	1	2	3	4	Total
No. ewes	38	117	139	25	2	321
% of total	11.8	36.4	43.3	7.8	0.6	100

Crosses With Wooled Breeds

Table 2.9.8 shows the fertility and litter size of purebred Blackbelly matings and pure and crossbred matings involving the imported Suffolks and Dorsets at the Greenland Station during 1977 and 1978. Clearly, the imported sheep demonstrated poor fertility in this environment, whether mated within breed or crossed with the Blackbelly; the low fertility rates of these exotics can be traced to both sexes. These preliminary indications are not encouraging for the fertility and prolificacy of these crossbred ewes. The performance of the Blackbelly breed was very good, particularly with regard to prolificacy.

Preliminary observations have been made on carcass traits of Blackbelly and crossbred lambs. Dressing percentages for the three groups (BB, D × BB, and S × BB) were virtually identical (Table 2.9.9). There was little variation among breed groups for individual carcass cuts (expressed as a percentage of carcass weight). The distribution of muscle (especially from the leg, where Blackbelly has been thought to be deficient) showed no marked differences between the breeds. It appears that the stocky types, Suffolk and

TABLE 2.9.8.   FERTILITY AND PROLIFICACY IN BLACKBELLY (BB), SUFFOLK (S), DORSET (D), AND SOME CROSSBRED MATINGS, GREENLAND STATION, 1977-78

Ram	Ewe	No. exposed	No. lambed	% lambed	Total	No. lambs born per ewe Exposed	No. lambs born per ewe Lambed
BB	BB	299	234	78.3	468	1.56	2.00
	S	15	8	53.3	11	0.73	1.38
	DxBB	11	4	36.0	4	0.36	1.00
	SxBB	3	2	66.7	2	0.66	1.00
D	BB	162	84	51.8	168	1.04	2.00
	D	53	3	5.7	5	0.09	1.67
S	BB	76	36	47.4	78	1.03	2.17
	S	52	13	25.0	15	0.29	1.15
By breed of ram							
All BB		328	248	75.6	485	1.48	1.96
All D		215	87	40.5	173	0.80	1.99
All S		128	49	38.3	93	0.73	1.90
By breed of ewe							
All BB		537	354	65.9	714	1.33	2.02
All D		53	3	5.7	5	0.09	1.67
All S		67	21	31.3	26	0.39	1.24

TABLE 2.9.9.   CARCASS TRAITS

Group	No.	Livewt. kg	Dressing %	% of cold carcass wt in Forequarter[a]	% of cold carcass wt in Leg + loin
Blackbelly rams	23	27	43	52	41
Dorset cross wethers	24	33	42	52	41
Suffolk cross wethers	9	31	43	55	43

[a] Includes breast, shank, flank, shoulder, and rack.

Dorset crosses, have shorter muscles and bone length; the same amount of muscling is stretched over a longer bone in the case of Barbados Blackbelly sheep.

ACKNOWLEDGMENTS

The author expresses appreciation for the assistance of Mr. Bruce Lauckner—Biometrician seconded to the Caribbean Agricultural Research and Development Institute from the British Ministry for Overseas

Development—who was responsible for the analysis of the data, and Mr. Anthony Sobers—Officer-in-Charge, Greenland Sheep Development Programme—who interpreted and updated records at Greenland.

REFERENCES

Combs, W. 1979. Personal communications.

Ligon, R. 1657. *A true and exact history of the island of Barbados*. London: Peter Parker and Thomas Grey.

Patterson, H. C. 1976. The Barbados Blackbelly sheep. 19 pp. Bulletin of Barbados Ministry of Agriculture, Science and Technology, Bridgetown, Barbados.

Patterson, H. C. 1978. The importance of Blackbelly sheep in regional agriculture. Mimeo. 9 pp. Paper presented at Second Regional Conf. on Sheep Prod., Bridgetown, Barbados, Sept. 1978.

REPRODUCTION AND BIRTH WEIGHT OF BARBADOS BLACKBELLY SHEEP IN THE GOLDEN GROVE FLOCK, BARBADOS

G. E. Bradford, *University of California, Davis, U.S.A.*
H. A. Fitzhugh, *Winrock International, U.S.A.*
A. Dowding, *Golden Grove Estates, Barbados*

SOURCE OF DATA

The Golden Grove Estate flock of Barbados Blackbelly sheep is privately owned and maintained as part of a mixed crop and livestock plantation operation.

The sheep described in this report were kept on pastures during the day and confined to a corral during the night. They grazed on forage species typical of the island, as well as on harvested fields of vegetable-crop, when available. The flock was well managed and in good condition. (Temperature and rainfall patterns for Barbados are described by Patterson in Chapter 2.9.)

More than 90% of the ewes in the flock had a typically Blackbelly phenotype; a few carried a 25% or less inheritance from the Devon Longwool breed and some had a small percentage of Wiltshire Horn ancestry. Litter-size data were obtained for two groups: (1) Blackbelly ewes and (2) crossbred ewes and ewes with some wool.

Lambing records kept from March 1975 to May 1978 provided information on litter size and lambing intervals from a total of 303 different ewes and 554 lambings. Many of the ewes were past first lambing when recording began; age and litter size at first lambing were known for only 22 ewes. Birth weights were recorded on 279 lambs. Perinatal mortalities were recorded for all 554 lambings.

Rams were with the ewes continuously, except for two periods of approximately 3 months each (March–May 1977 and July–December 1977). A small number of matings occurred during these two periods, probably by

TABLE 2.10.1.   LITTER SIZE AT BIRTH

Group	Lambing sequence	No. ewes	No. lambings	No. ewes with					$\bar{x}$	SD
				1	2	3	4	5		
Black-belly	First	20	20	12	7	1	–	–	1.45	.60
	Second & later	125	216	75	94	43	4	–	1.89	.78
	All	273	489	170	229	78	10	2	1.86	.78
	% of total			34.8	46.9	15.8	2.1	0.4		
Cross-bred[a]	All	30	65	24	33	8	–	–	1.75	.66

[a] Ewes carrying 25% or less of inheritance from Devon Longwool or Wiltshire Horn (remainder Blackbelly) and ewes with any wool.

unweaned ram lambs, but lambings from August to October 1977 and from December 1977 to February 1978 were less than 10% of those in the periods that immediately preceded and followed. Lambing intervals were analyzed for: (1) all data, and (2) data from all intervals completed on or before July 31, 1977.

PERFORMANCE

Litter Size

Table 2.10.1 summarizes the mean litter size per ewe. Means and distributions are given for known first lambings, for known second and later lambings, and for all lambings, including unknown parity. The few ewes known to be lambing for the first time had appreciably smaller litters (1.45). Crossbred ewes, including those recorded as having some wool, had slightly smaller litters (1.75) than did typical Blackbelly ewes (1.86).

The mean litter size of 1.86 for all Blackbelly ewes was slightly lower than the 2.00 reported by Patterson (Chapter 2.9) for another flock in Barbados, but the pattern of variation in the two flocks was quite similar. Patterson reported 28% with 1 lamb, 48% with 2, 21% with 3, 3% with 4, and .1% with 5. These data indicate considerably more variation in litter size for the Blackbelly breed than that for other breeds of sheep. Suffolk sheep had a mean litter size of 1.85 (330 litters) in the University of California, Davis flock (Bradford, unpublished). The mean reported by Bradford is almost identical to that obtained in this study, but the Suffolk sheep litter percent-

TABLE 2.10.2.   CORRELATIONS BETWEEN NUMBER OF LAMBS
                  BORN IN DIFFERENT LITTERS OF THE SAME EWES

Litter sequence	Litter sequence								
	1[a]			2			3		
	No.	r	SE	No.	r	SE	No.	r	SE
2	145	.13	.13						
3	64	.23	.07	64	−.13	.30			
4	28	.23	.23	28	−.01	.97	28	.01	.96

[a] First recorded litter; may not have been ewe's first litter.

ages were: 1 lamb, 29%; 2 lambs, 58%; 3 or more lambs, 13%, which compares with 1 lamb, 35%; 2 lambs, 47%; and 3 lambs or more, 18% for this study. The coefficient of variation (CV) of 42% for the data in Table 2.10.1 can be compared with the average CV of 38% for 7 breeds of sheep reported by Glimp (1971) and a CV of 28% for 297 Finnish Landrace litters in the ABRO flock reported by Bradford et al. (1974). The Booroola Merino and its crosses (Allison et al. 1978) are the only sheep reported to have a coefficient of variation in litter size similar to that of the Blackbelly.

The repeatability of litter size is of interest, particularly in light of this high variability. Spearman correlation coefficients were calculated using data from 2 to 4 lambings. The "first" litters in this sample were the first recorded litters, but not necessarily litters from maiden ewes. The results are summarized in Table 2.10.2.

The correlations between number born in the first recorded litter and each of three subsequent litters were positive and are in reasonable agreement with published estimates of repeatability for litter size (Turner 1969). Correlations between number of lambs in adjacent litters were less than those between nonadjacent litters. Adjacent litters would show negative correlations if a larger litter resulted in the ewe being in below-average condition at next breeding, and vice versa; this might be expected to be a more important factor with a 7- to 8-month lambing interval than with a 1-year interval. The correlation between first and second recorded lambings (1 × 2) was lower than that for 1 × 3 or 1 × 4. The correlation for 2 × 3 was lower than that for 2 × 4, which is consistent with the hypothesis suggesting a negative correlation due to environment between adjacent litters.

Table 2.10.3 shows the effect of season on litter size. The mean litter sizes for the four calendar years were: 1975, 1.87; 1976, 1.86; 1977, 1.84; 1978, 1.87. In the absence of evidence suggesting differences due to year, the data for each month were pooled across years. The means for March, April, and May lambings were based on 4 years' data, with those from the remaining months based on 3 years' data.

TABLE 2.10.3.   MEAN LITTER SIZE BY MONTH OR SEASON
                     OF LAMBING

Month	No. lambings	Mean litter size	Seasonal mean	Mating dates
March	69	1.81		
April	45	1.80	1.83	Oct–Dec
May	76	1.86		
June	41	1.73		
July	29	1.90	1.73	Jan–Mar
August	24	1.54		
September	17	2.11		
October	50	2.06	1.96	Apr–June
November	110	1.89		
December	31	1.84		
January	38	1.76	1.82	July–Oct
February	24	1.88		

The smallest seasonal litter size followed matings during the dry season (January through March), with the largest size following matings during the earlier part of the wet season and probably representing a flushing effect. There was no evidence of reduced litter size due to "out-of-season" matings, which is characteristic of seasonally fertile breeds including those with a relatively long breeding season.

Age at First Lambing

The mean age for 22 ewes at first lambing was 467 days, with a range from 329 to 730. (If the ewe that lambed at 730 days were excluded, the mean would be 454 and the maximum 519 days.) The modal age class was 470 to 500 days, or 16 months.

Lambing Dates and Lambing Interval

Lambings occurred in every month during which records were kept, although there were few lambings during two 3-month intervals between 1975 and 1978 or as a result of a deliberate attempt to group lambings (as explained earlier).

The mean lambing interval for all ewes was 253.3 days. For the 151 intervals completed before July 31, 1977 (Group I), the mean was 244 days, with a standard deviation of 76 days and a range of from 158 to 520 days. For the 100 intervals completed after July 31, 1977 (Group II), the mean was 267 days, an increase that could be accounted for by breeding-flock manage-

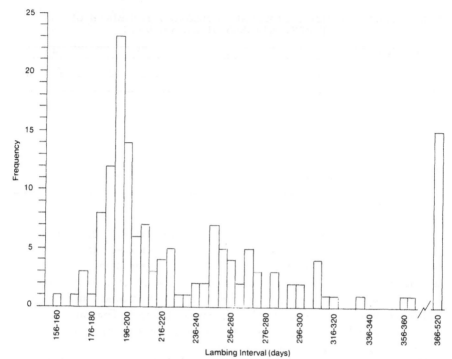

FIGURE 2.10.1 Frequency distribution of lambing intervals in a flock of Blackbelly ewes exposed continuously to rams.

ment. The mean interval in this flock is similar to the 257 days reported by Patterson (Chapter 2.9) for an experimental flock of Blackbelly ewes.

The intervals for pure Blackbelly ewes were almost identical to those of ewes carrying some inheritance of other breeds; thus the data presented are for all ewes in the flock.

The distribution of lambing intervals for Group I ewes is shown in Figure 2.10.1. The distribution was markedly skewed, as expected for a trait with a biologically fixed lower limit that can be extended at the upper end by a variety of environmental factors.

The minimum interval of 158 days was somewhat surprising because ewes rarely conceive sooner than 30 days postpartum. (No error was indicated, however—the interval followed the birth of a single lamb in October, and so the value was included.)

There was a distinct mode of intervals at 190 to 195 days, indicating a normal postpartum anestrus for these sheep of 40 to 45 days, with a "normal" range of 20 to 70 days. A small secondary peak occurred at around 250 days.

TABLE 2.10.4.   LAMBING INTERVALS FOLLOWING THE BIRTH OF
                 DIFFERENT NUMBERS OF LAMBS, DAYS

Litter size	1			2			3			4		
Group	No.	$\bar{x}$	SD	No.	$\bar{x}$	SD	No.	$\bar{x}$	SD	No.	$\bar{x}$	SD
Ewes completing intervals before July 31, 1977	50	212.8	36.7	62	233.6	46.2	19	215.8	35.8	4	206.2	36.5
Ewes completing intervals after July 31, 1977	28	237.2	52.6	36	255.8	53.7	26	243.5	49.0			

The observed skewed distribution in lambing intervals could be due to a skewed distribution in date of first estrus (i.e., to a seasonal delay in puberty or postparturition estrus for some ewes), longer lactational anestrus associated with larger litters, or other causes. The distribution of lambing intervals also reflects factors such as ewes not being mated when in estrus and conception failures due to either the ewe or the ram. The data from this flock did not allow evaluation of all these possibilities, but the effects of season and of litter size were examined.

There were differences in lambing intervals between seasons that were significant in a few cases, but these differences were not consistent from year to year. Overall, there were no discernible seasonal effects on lambing interval.

Lambing intervals following birth of 1, 2, 3, and 4 lambs are shown in Table 2.10.4. Intervals exceeding 365 days were excluded. In both groups of ewes, the interval following birth of twins was longer than that following birth of singles; this difference was significant ($P < 0.05$) for the first group. Such a difference would be expected if length of postpartum anestrus were positively correlated with level of lactation stress. Surprisingly, ewes with triplets (and the few with quadruplets) had *shorter* lambing intervals than did ewes with twins; in fact, their intervals were similar in length to those of ewes with singles, for both groups.

Birth Weights

Birth weights were recorded for 279 lambs from this flock (Table 2.10.5). All lambs were included, since all were sired by Blackbelly rams and were at least 7/8 Blackbelly breeding.

Unfortunately, there were no recorded mature ewe weights from this flock. However, the mean mature weight was estimated at 40 to 45 kg. Based on this estimate the predicted birth weight of singles, using the formula of Donald and Russell (1970), would be between 3.34 and 3.64 kg. The

TABLE 2.10.5.   BIRTH WEIGHTS OF BLACKBELLY LAMBS BY SEX AND LITTER SIZE, kg

	Males			Females				
Size	No.	$\bar{x}$	SD	No.	$\bar{x}$	SD	Avg wt	Ratio[a]
1	20	3.72	.91	29	3.32	.76	3.52	
2	89	2.89	.37	59	2.86	.31	2.88	0.82
3	31	2.94	.49	39	2.58	.59	2.76	0.78
4	4	2.61	.39	8	2.47	.23	2.54	0.72

[a] Avg. weight of multiples/avg. weight of singles.

TABLE 2.10.6.   LAMB MORTALITY BY SEX AND LITTER SIZE

No. born in litter	Males			Females			Total		
	Born	Died	%	Born	Died	%	Born	Died	%
1	114	5	4.4	80	4	5.0	194	9	4.6
2	270	19	7.0	254	15	5.9	524	34	6.5
3	123	26	21.1	135	26	19.2	258	52	20.2
4	18	6	33.3	22	7	31.8	40	13	32.5
5	6	3	50.0	4	1	25.0	10	4	40.0
Total	531	59	11.2	495	53	10.7	1026	112	10.9

observed mean for singles is within the predicted range. However, the weights of multiples are higher than predicted by the formula of Donald and Russell (1970), which predicts weights of twins and triplets, or that of Dickinson et al. (1962), which predicts weights of lambs in any litter size. Donald and Russell suggested that the curve relating birth weights to litter size for the Finnish Landrace differs from that for other breeds, with a less-than-expected decline with increasing litter size. The data in Table 2.10.3 suggest this same pattern for the Blackbelly breed, and it is possible that this flatter slope is characteristic of "prolific" breeds.

Lamb Survival

Records were kept on lambs born dead or that died during the first few weeks after birth. These data provide a basis for unbiased comparisons of relative mortality for males and females and for lambs from different litter sizes (Table 2.10.6).

Mortality rates for the two sexes were similar for all litter sizes. Mortality was low for both singles and twins. Although mortality increased with litter size, the number of viable lambs in litters of 3, 4, and 5 confirm the potential advantages of high prolificacy under good postnatal management.

SUMMARY

A commercial flock of Barbados Blackbelly ewes in Barbados had a mean litter size of 1.86 based on 489 lambings. The litter size percentages were: 34.8%, 1 lamb; 46.9%, 2; 15.8%, 3; 2.1%, 4; and 0.4%, 5. Lambings occurred throughout the year. Litter size was 1.73 following matings in January through March (dry season) and 1.96 following matings in April through June. Age at first lambing varied from 11 to 24 months, with a modal class of 16 months.

A mean lambing interval of 244 days was obtained over 151 lambing intervals, during which the ewes were exposed continuously to rams. The distribution was highly skewed, with a modal class of 190 to 195 days and a range of 158 to 520 days. Exclusion of ewes with an interval greater than 365 days reduced the mean to 223 days, with a standard deviation of 42.4 days. Intervals following birth of twins were about 20 days longer than those following birth of singles, and also longer than those following birth of triplets.

Birth weight of single lambs averaged 3.5 kg. The birth weight of a twin averaged 82% of a single; a triplet, 78%; a quadruplet, 72%. These data represent a lesser effect of litter size on birth weight than that reported for most breeds of wool sheep.

The lamb mortality rate averaged 11% for both male and female lambs: the mortality rate for single lambs was 4.6%; for twins, 6.5%; for triplets, 20.2%; for quadruplets, 32.5%; and for quintuplets, 40.0%.

REFERENCES

Allison, A. J., J. R. Stevenson, and R. W. Kelly. 1978. Reproduction and wool production of progeny from high fecundity (Booroola) Merino rams crossed with Merino and Romney ewes. World Conf. Anim. Prod., Buenos Aires.

Bradford, G. E., St.C.S. Taylor, J. F. Quirke, and R. Hart. 1974. An egg-transfer study of litter size, birth weight and lamb survival. *Anim. Prod.* 18:249-263.

Dickinson, A. G., J. L. Hancock, G.J.R. Hovell, St.C.S. Taylor, and G. Wiener. 1962. The size of lambs at birth—a study involving egg transfer. *Anim. Prod.* 4:64-79.

Donald, H. P. and W. S. Russell. 1970. The relationship between live weight of ewe at mating and weight of newborn lamb. *Anim. Prod.* 12:273-280.

Glimp, H. A. 1971. Effect of breed and mating season on reproductive performance of sheep. *J. Anim. Sci.* 32:1176-1182.

Turner, H. N. 1969. Genetic improvement of reproduction rate in sheep. *Anim. Breed. Abstr.* 37:545.

VIRGIN ISLANDS WHITE HAIR SHEEP

Harold Hupp, *Virgin Islands Agricultural Experiment Station, U.S.A.*
Duke Deller, *Virgin Islands Department of Agriculture, U.S.A.*

INTRODUCTION

Sheep flocks of the Virgin Islands are usually small, cared for by family members, and serve as a source of supplemental income. The local population enjoys mutton, and recently the islands have shown a slight increase in sheep numbers. The 1975 Agricultural Census lists 3,122 head of sheep on 64 farms, an increase from 2,185 head on 31 farms in 1970, and from 2,152 head on 93 farms in 1960. The majority of the flocks have 100 head or fewer.

LOCATION AND CLIMATE

The U.S. Virgin Islands are approximately 60 km east of Puerto Rico, at 18° N and 65° W. Rainfall varies, depending primarily on island topography and the continual easterly tradewinds; the eastern portion of the islands receives only 500 to 760 mm annually, while the western portion receives 1,000 to 1,500 mm. December through April are the drier months of the year, with mean monthly precipitation ranging from 40 to 90 mm. The range in means is 90 to 150 mm per month for the months of May to November. Mean monthly daytime temperature varies from a low of 28° C in January and February to a high of 32° C in July to September. Nights average 7° C cooler than days, and daylight varies from a low of 11.18 hours in January to a high of 13.14 hours in June.

FEED RESOURCES

Poor nutrition is one of the main factors limiting sheep production. Pastures vary in quality from poor to fair. Pastures are primarily hurricane grass, buffel grass, and tan-tan (*Leucaena*). The improved pastures consist of

Guinea grass and Pangola grass. The major problems are the fluctuation and scarcity of the pasture quality and quantity, the high cost of labor and of imported concentrate feeds, and the lack of farm machinery. Feed shortages occur during the dry season and in droughts. Some shepherds hand-cut grass to supplement the diet of their livestock. Little concentrate feed is used, except at critical times such as the dry season.

MANAGEMENT

Tethering and extensive management practices are used, with sheep often raised along with goats. Sheep range freely, or graze fenced pastures with no systematic rotational grazing. The pastures are often old, poorly maintained sugarcane fields. Animals generally are maintained on a low level of nutrition.

The flocks are managed as a unit with little or no division into groups by age and sex. The animals are corralled at night, since dog attacks and pilferage problems are very serious. At lambing, ewes and young lambs are corralled for 1 to 2 weeks until lambs are strong enough to go to the pasture. Rams remain with the flock continuously.

Rams are exchanged among flocks periodically as a means of limiting inbreeding, but most female replacements are selected from the owner's flock. Ram lambs are usually castrated at 2 to 4 months of age. Lambs are wormed at 2 to 4 months and sometimes again at 6 to 8 months, while mature animals are usually wormed once or twice a year.

MARKETING

Sales of live animals are private; producers deal directly with individuals or restaurants. Occasionally, when the supply exceeds the private demand, the excess animals are sold to the local meat markets at a reduced price.

ORIGIN

The origin of the Virgin Islands White Hair Sheep is not clear. Bond (1975) and Mason (1980) suggest that these sheep came from West Africa, noting their similarity to other sheep in the Caribbean area for which a West African origin is generally accepted. Another theory holds that the Wiltshire Horn may have been imported and crossed with the native creole sheep (Faulkner 1962; Devendra 1977). An attempt is being made to trace the origin of these sheep through shipping logs in the Danish archives. Little selection has been practiced by producers. The occasional introduction of temperate and tropical breeds has further compounded the development of these sheep. Older people on the islands say that the sheep of today are smaller and less prolific than they were earlier in this century.

DESCRIPTION AND SPECIAL CHARACTERISTICS

The White Hair sheep is a small breed with fairly good mutton characteristics. Both sexes are polled. "Pure" sheep have a white head and body, light short hair that is longer over the back and rump, and pink skin with occasional dark spots; occasionally they have some wool on the hind quarters. Some sheep are solid tan, brown, black, or white with brown or black spots. Ears are medium length, carried horizontally, and often pointed forward. The tail hangs down to the hocks. The tarsus of the eye is black with dark eyelashes, while the iris of the eye is amber. The feet are of ample size with light-gray hooves. Mature rams have a well-developed mane.

These sheep are believed to be less susceptible to internal parasites than are goats and seem to be unaffected by ticks. Lumpy jaw and helminth worms are the major health problems on the Island of St. Croix.

PERFORMANCE

Weight, measurements, and information on reproduction were recorded in the period January to April 1979 on sheep in three privately owned flocks with 30, 60, and 187 sheep per flock; two flocks were on St. Croix, the other on St. Thomas. Management in these flocks was as described previously (except that goats were not run with the sheep in two of the three flocks).

Data were initially summarized separately for pure white sheep ("purebred") and for colored or spotted sheep or sheep with some wool ("crossbred"). Mean weights and body measurements were virtually identical for the two groups, in all age and sex classes; thus the data were pooled as shown in Table 2.11.1. Lamb data were not adjusted for age, as birth dates were not known.

The data indicate a mature weight of about 35 kg for lactating ewes during the drier part of the year. The only two mature rams weighed 37 and 54 kg. Rams of this breed with weights as high as 90 kg have been recorded on the Island of St. Croix, but such weights are rare.

Numbers of lambs born per ewe were recorded in two flocks (Table 2.11.2). The mean number of lambs born in the two flocks differed substantially, but it is not known whether this is due to genetic or nutritional factors, or a combination of the two. The breed, in general, has a reputation among flock owners of being prolific, with reported multiple-birth rates typically in the range of 50 to 75%. The data shown are in reasonable agreement. Ewes are known to lamb more often than once a year, but no data are available on actual lambing intervals. The description of these sheep by

TABLE 2.11.1.   BODY WEIGHTS AND MEASUREMENTS OF VIRGIN ISLANDS (ST. CROIX AND ST. THOMAS) WHITE HAIR SHEEP

Age[a]	Sex	Flock	Weight, kg			Body measurements, cm						
							Height at withers		Heart girth		Length[b]	
			No.	x̄	SD	No.	x̄	SD	x̄	SD	x̄	SD
Lambs	M	1	45	7.7	1.8	38	43.6	3.0	44.7	4.5	32.8	3.5
		2	6	8.1	2.5	6	42.3	3.8	41.5	8.6	25.8	4.1
		3	3	6.5	3.0	3	39.9	3.0	41.1	6.4	26.2	5.8
	F	1	44	6.8	1.5	38	42.6	3.2	42.1	3.9	31.4	3.1
		2	14	8.6	2.8	14	41.6	6.2	43.8	8.8	26.1	4.9
		3	2	10.7	4.9	2	43.9	6.4	49.5	5.3	27.4	4.6
6–12 mo.	wethers	3	6	20.5	1.9	6	56.4	3.0	62.0	3.6	36.1	1.0
	F	1	17	23.2	2.8	17	59.9	9.4	66.0	3.6	41.7	3.6
		3	2	19.5	0.6	2	54.6	1.8	63.5	3.6	37.6	2.8
Mature	F	1	78	32.6	5.9	78	64.3	3.4	75.7	5.3	45.4	3.0
		2	38	33.4	7.0	38	62.5	3.4	76.7	6.5	37.8	2.1
		3	18	37.4	5.6	18	63.0	3.3	79.0	5.1	42.9	4.1

a Lambs were 2 to 4 months of age.

b Measured from the top of the shoulders to the top of the hip.

TABLE 2.11.2. LITTER SIZE IN TWO FLOCKS OF VIRGIN ISLANDS (ST. CROIX) WHITE HAIR SHEEP

Flock no.	Period recorded	No. ewes	No. lambs		Mean litter size	
			Born	Weaned[a]	Born	Weaned
1	Feb–April	54	78	68	1.44	1.26
3	Jan–March, Aug–Dec	19	35	32	1.84	1.68

[a] Of the total loss of 13 lambs, 7 were stillborn and 6 were lost due to other causes.

Bond (1975) as having 2, 3, or 4 lambs per lambing and 2 lambings per year may overestimate both traits, at least under local management conditions.

SUMMARY

The Virgin Islands White Hair Sheep are of modest size, with ewe and ram weights of approximately 35 and 45 kg, respectively, under typical commercial conditions. Litter size averaged 1.44 and 1.84 in two flocks, supporting the belief that the breed is prolific. These sheep appear to be well-adapted to the warm humid conditions of the Caribbean. Limitations in quantity and quality of pasture and lack of an organized marketing system are important constraints to production in the Virgin Islands at this time.

ACKNOWLEDGMENTS

The authors are indebted to Mr. Arthur Hartman, St. Thomas, and Mr. William Johansen and Mr. Axel Fredriksen, St. Croix, for allowing the authors to collect data on their flocks.

REFERENCES

Bond, R. M. 1975. Small livestock for meat production in the Virgin Islands. Virgin Islands Agriculture and Food Fair Bulletin.

Devendra, C. 1977. Sheep of the West Indies. *World Review of Animal Production.* 13(1):31–38.

Faulkner, D. E. 1962. Report on livestock development in the British Virgin Islands. Interim Commission of the West Indies, Port of Spain, Trinidad.

Mason, I. L. 1980. Prolific tropical sheep. FAO Animal Production and Health Paper 17. Rome: FAO.

Park, W. L. and R. L. Park. 1974. Potential returns from goat and sheep enterprises in the U.S. Virgin Islands. CVI Agricultural Experiment Station Report No. 7.

A NOTE ON PERFORMANCE OF
BARBADOS BLACKBELLY SHEEP IN JAMAICA

G. E. Bradford, *University of California, Davis, U.S.A.*
A. J. Muschette, Vincent Lyttle, and David Miller,
Agricultural Development Corporation, Jamaica

The Agricultural Development Corporation, Hounslow, maintained a flock of Barbados Blackbelly ewes which provided the reproduction data shown in Table 2.12.1.

Other data were available from the private flock of Mr. Pat McDaniel, Kingston, which consisted of 18 high-grade Blackbelly ewes maintained with high-quality nutrition throughout the year. The farm elevation is about 1000 m.

A total of 81 lambings had been recorded for the 18 ewes, although 7 of the ewes were 36 months old or less at the time the data were summarized. One ewe had 10 lambings in a period of 60 months, which was the maximum lambings per ewe reported. Another ewe had 14 lambs in 5 litters over a period of 27 months.

Data from this flock are summarized in Table 2.12.2.

The differences in performance between the two above flocks may have been caused by several factors, including level of concentrate feeding, elevation, and flock size. The editors have a subjective, but very definite, impression formed in observing many flocks: that Barbados Blackbelly ewes kept in small flocks perform better than do those in large groups. Unfortunately, ewe weights were not available for either flock.

A limited amount of data also was available on the St. Elizabeth breed, Jamaica's "native" sheep, which are a medium-small, wooled sheep with a twinning rate of less than 10%. They appear to be much better adapted to the climatic conditions of Jamaica than are recently imported wool sheep. The breed has been described by Mason (1978). No data on these wooled sheep are presented here; however, the seasonal breeding pattern of this breed, which has evolved for many years in a tropical climate, was com-

TABLE 2.12.1.   LAMBING PERFORMANCE OF BARBADOS BLACK-
BELLY EWES BY SEASON. HOUNSLOW, JAMAICA

Season[a]	No. ewes lambed	% of total	Litter size $\bar{x}$	SD	No. ewes with 1	2	3
Jan.-Mar.	71	19.4	1.30	.46	50	21	-
Apr.-June	108	29.6	1.47	.59	62	41	5
July-Sept.	40	11.0	1.30	.56	30	8	2
Oct.-Dec.	146	40.0	1.41	.53	89	54	3
Total	365	100.0	1.39	.54	231	124	10

Lambing interval, days $\bar{x}$ 252.9, SD 30.6

[a] Lambing occurred in every month of the year.

TABLE 2.12.2.   PERFORMANCE OF GRADE BARBADOS BLACKBELLY
EWES IN THE McDANIEL FLOCK, JAMAICA

Trait	No. obs.	$\bar{x}$	SD	No. ewes with 1	2	3	4
No. born, 1st litters	18	1.72	.67	7	9	2	-
No. born, 2nd + litters	63	2.05	.85	17	30	12	4
Age at first lambing, days	16	396.4	82.7	(range 286-581)			
Lambing interval, days	63	228.4	42.6	(range 169-331)			

pared with that of hair sheep. Hare (1977) reported the results of checking estrus daily for more than a year on 11 two- and three-year-old St. Elizabeth ewes fed in confinement. First estrus of the season was observed as follows: April, 1 ewe; May, 2 ewes, June, none; July, 7 ewes; August, 1 ewe. Mean days from first to last estrus for 8 ewes was 188 days with a range of 146 to 236 days and a standard deviation of 29 days. Three ewes that were bred accidentally and lambed in November did not return to estrus before the end of the breeding season. Thus, the seasonal pattern shown by the St. Elizabeth ewes was like that of temperate breeds and quite different from that of hair sheep in this environment.

REFERENCES

Hare, L. 1977. A preliminary study of the reproductive potential of the St. Elizabeth sheep. Mimeo. 6 pp. Ministry of Agriculture, Jamaica.

Mason, I. L. 1978. Conservation of animal genetic resources. Prolific tropical sheep. FAO/UNDP. Mimeo. 22 pp. Rome: FAO.

A HISTORY OF THE BARBADOS BLACKBELLY SHEEP

Weslie Combs, *Ottawa, Ontario, Canada*

Barbados Blackbelly sheep combine the rare attributes of adaptation to tropical environments and high reproductive efficiency, which account for their average of two lambs per litter and an average lambing interval of 8 to 9 months (Patterson 1978).

Several writers (Devendra 1972, Patterson 1976, Maule 1977, and Mason 1980) have speculated on the origin of the breed, which has been widely accepted as African. Although there can be little doubt that the Blackbelly has African ancestry, there is compelling historical evidence that the Barbados Blackbelly as a breed originated and evolved on the island of Barbados. Following the colonization of Barbados by the English in 1627, several fortuitious factors existed in combinations that may not have occurred in other early European colonies in the Americas. These factors were:

1. The introduction of tropically adapted hair sheep with the slave trade from West Africa. There is, of course, good reason to believe that African sheep were taken independently to other colonies in the West Indies and Latin America. In combination with other stocks, various local types and breeds were founded, possibly including wooled Criollo breeds.
2. The concurrent introduction to Barbados of a poorly adapted, but prolific, wooled sheep of unknown origin. The most likely country of origin may have been Holland or England, as each had active colonization and commerce in the region. This unknown parent breed may, more than any other factor, set the Barbados Blackbelly apart genetically from other breeds in the region.
3. An environment, including heat, humidity, and parasites, to which wooled sheep were not well adapted. Of particular difficulty were

the small burrs in the pastures that clung to and penetrated the fleece; they may have so afflicted the wooled sheep as to have placed constant selective pressure against woolliness. This factor could have kept the evolving gene complex of the Blackbelly from being overwhelmed by that of wooled breeds imported from time to time to "improve" the mutton characteristics of the Barbados sheep. However, selection against woolliness would not have precluded introgression of other genes such as those for prolificacy or polledness into the Blackbelly.

4. Sugarcane culture that rapidly forced livestock into the roadsides and marginal lands where burrs abound and into a system of management that provided the opportunity for a measure of control of matings and for artificial selection.

THE AFRICAN ANCESTOR

The classic work on the early history of Barbados is the remarkable record by Richard Ligon (1657), who lived and worked on a plantation in Barbados from 1647 to 1650. The book was written about 1653 while Ligon was a prisoner in Upper Bench prison in London, apparently for debts contracted before going to Barbados. Publication costs were underwritten by the Bishop of Salisbury who had urged Ligon to begin the book.

Ligon's shipmate from England to Barbados, Col. Thomas Modiford (Modyford), had bought half interest in the Hilliard plantation, today known as Kendall, and had made Ligon his assistant. Modiford later became Governor of Jamaica, and the extracts from his dispatches in the Colonial Papers provide further background on sheep in the region.

In emphasizing the African descent of the "Barbados woolless sheep," Buttenshaw (1906) only partially quoted Ligon's commentary on the early sheep of Barbados and omitted Ligon's reference to wooled sheep.

Subsequent histories of the Blackbelly have perpetuated this oversight. Ligon's full statement was (first edition, written 1653, published 1657):

> Sheepe. We have here, but very few; and those do not like well the pasture, being very unfit for them; a soure tough and saplesse grasse, and some poysonous plant they find, which breeds diseases amongst them, and so they dye away, they never are fat, and we thought a while the reason had been, their too much heate with their wool, and so got them often shorne; but that would not cure them: yet the Ews bear alwayes two Lambs: their flesh when we tried any of them, had a very faint taste, so that I do not think they are fit to be bred or kept in that Countrey: other sheep we have there, which are brought from Guinny and Binny, and those have haire growing on them, instead of wool; and liker Goates then Sheep, yet their flesh is tasted more like Mutton then the other.

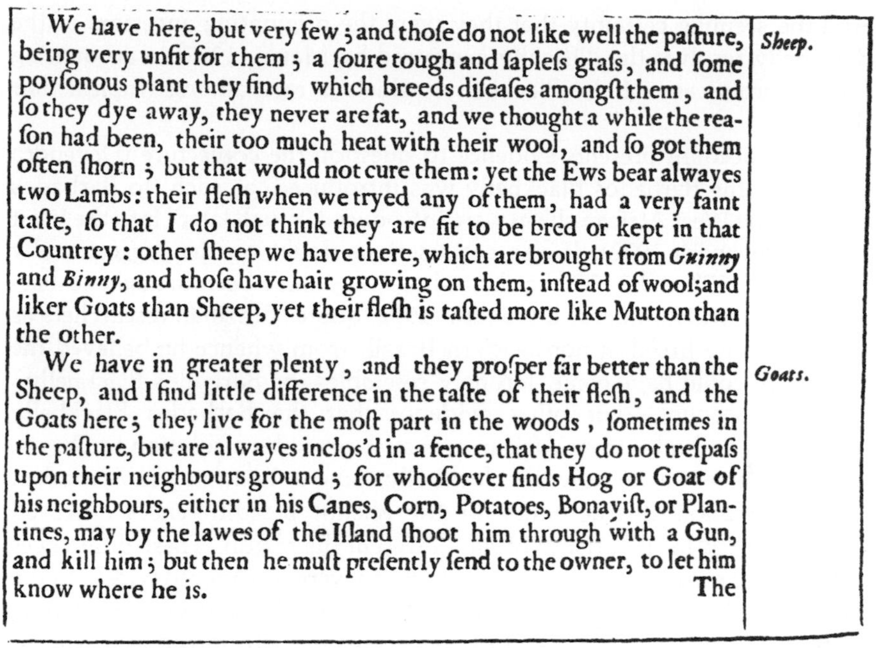

We have here, but very few ; and thofe do not like well the pafture, | *Sheep.*
being very unfit for them ; a foure tough and faplefs grafs, and fome
poyfonous plant they find, which breeds difeafes amongft them, and
fo they dye away, they never are fat, and we thought a while the rea-
fon had been, their too much heat with their wool, and fo got them
often fhorn ; but that would not cure them: yet the Ews bear alwayes
two Lambs: their flefh when we tryed any of them, had a very faint
tafte, fo that I do not think they are fit to be bred or kept in that
Countrey : other fheep we have there, which are brought from *Guinny*
and *Binny*, and thofe have hair growing on them, inftead of wool; and
liker Goats than Sheep, yet their flefh is tafted more like Mutton than
the other.

We have in greater plenty, and they profper far better than the | *Goats.*
Sheep, and I find little difference in the tafte of their flefh, and the
Goats here ; they live for the moft part in the woods , fometimes in
the pafture, but are alwayes inclos'd in a fence, that they do not trefpafs
upon their neighbours ground ; for whofoever finds Hog or Goat of
his neighbours, either in his Canes, Corn, Potatoes, Bonavift, or Plantines, may by the lawes of the Ifland fhoot him through with a Gun,
and kill him ; but then he muft prefently fend to the owner, to let him
know where he is. The

FIGURE 2.13.1 **Photograph of page from Ligon (second edition, 1673) with Ligon's description of the sheep and goats of Barbados as of 1650.**

Goates. We have in greater plenty, and they prosper farre better than the Sheep, and I find little difference in the taste of their flesh, and the Goats here; they live for the most part in the woods, sometimes in the pastures, but are alwayes inclosd in a fence, that they do not trespass upon their neighbours ground; for whosoever finds Hog or Goat of his neighbours, either in his Canes, Corne, Potatoes, Bonavist, or Plantines, may by the lawes of the Island shoot him through with a Gun, and kill him; but then he must presently send to the owner, to let him know where he is.

The second edition (Ligon 1673) "modernizes" the spelling and corrects the third from last word in the first paragraph from "then" to "than" (Figure 2.13.1).

Clearly, within the first quarter century of colonization, wooled sheep of unstated origin and hair sheep from Africa were found on Barbados. The wooled sheep were distressed by the tropical environment, but "the Ews bear alwayes two Lambs." The hair sheep from Africa were, by inference, adapted to the tropics and were said to have tastier flesh than the wooled sheep. Ligon, a chef and gourmet, should be taken as an expert witness on the eating qualities of mutton. These two diverse types of sheep could easily account for all of the characteristics of the modern Barbados Blackbelly.

And the seeming certainty that they were the originating ancestors of the Blackbelly resolves the difficulty expressed by Maule (1977) in accounting for the prolificacy of the Blackbelly, which is rare among tropical hair sheep.

Ligon's account presents evidence to question the conjecture (Devendra 1977) that the Barbados Blackbelly was introduced, as such, to Barbados and that the West African sheep were introduced "much later" into the West Indies (Devendra 1972). It also casts serious doubt on the suggestion by Devendra (1972 and 1977) that hair sheep may have been brought to Barbados from Brazil. Indeed, Devendra (1972) quotes Brazilian researchers as saying that no breed in northeastern Brazil, from whence he believed the Blackbelly could have come, strongly resembles the Barbados Blackbelly.

Ligon's account, plus other evidence that the Barbados Blackbelly evolved on Barbados from African ancestry, provides a solution to Devendra's puzzlement as to how the Blackbelly came to be found on Barbados. This explanation also accounts for the phenotypic differences observed among various local varieties of hair sheep in the region. These differences offer further evidence that African hair sheep were introduced widely in the region in the century before the settlement of Barbados, spawning numerous local hairy types and perhaps crossing with introduced wooled sheep to establish Criollo races.

Ligon identifies the source of the African hair sheep as "Guinny" and "Binny." The term "Guinny" probably represents the environs of the Gulf of Guinea extending from modern Ghana to Equatorial Guinea. The term "Binny" probably refers to the area around Benin, Nigeria, locale of the Bini tribe (Barnhart 1954) from which slaves were brought to Barbados (Hoyos 1978).

This is also the region of the Cameroon, origin of the "Cameroon sheep." Specimens in the Hellabrun Tiergarten, Munich, Germany, are smaller than the Barbados Blackbelly but have the same distinguishing color pattern (with black bellies), although they are of a darker and more uniform red. The males are horned, whereas only scurs and rudimentary horns occur with any great frequency in the Barbados Blackbelly. The possibility was suggested by the Tiergarten director (personal interview 1978) that Barbados Blackbelly sheep may have been introduced to the Tiergarten, but no record or knowledge of exportation to Germany was found in Barbados. Twinning among the Cameroon sheep at Hellabrun was estimated to be "no more than 8 to 10%." [Ed. note: I. L. Mason was given a figure of 20% in an earlier interview with the director of the Hellabrun Tiergarten.]

THE WOOLED ANCESTOR

During the first 25 years of colonization, Barbados established commerce throughout the West Indies, with the New England colonies, and with European countries including Russia. Wooled sheep may have been imported from these trading partners and other colonizing countries of Europe.

Barbados was one of the last of the islands of the West Indies to be settled by Europeans and the first settlers may not have brought their own sheep with them. Being the most eastward of the West Indies and lying in the Atlantic Ocean rather than in the Caribbean archipelago of the Lesser Antilles (Windward Islands), Barbados was off the primary shipping routes. The Arawak Indians had been forced from Barbados by the Carib Indians about 1200 A.D. Later, the Spanish had taken all of the Carib Indians from the island as slaves. When the Portuguese arrived in 1536, they found Barbados without human occupants. The Portuguese released swine on the island as a future source of provisions.

By 1625, when the first English ship arrived (a result of navigational error), Barbados had reverted to its primitive, overgrown state. An expedition marched along the west coast and claimed Barbados for England. The captain of this ship returned in 1627 with the first English settlers, who probably knew that the island had no apparent grazing land but was abundantly supplied with feral pigs. Indeed, as land was cleared, "body and soul" were kept together by hunting pigs in the nearby woods (Hoyos 1978) and the feral pigs were eventually hunted to extinction.

The leader of the Barbados colony visited the Dutch governor of Guyana and returned with seeds and with Amerindians to teach the English tropical agriculture. There was no mention of livestock. By 1629 more than 1,800 persons had settled on Barbados. They could have brought sheep; however, the only known pasturage was in the drier eastern side of the island where the land was very rough. It was the last to be settled. Cotton had been established, so wool was not a local necessity.

Volume I of the extracts of the Colonial Papers (Sainsbury 1860) that cover the first years of settlement shows no record of petitions to bring sheep from England to any of the English colonies. However, in Volumes II through IV (Sainsbury 1880–1893) several references are made to sheep established in English colonies other than Barbados.

Much to the displeasure of England, a major shift in commerce occurred in Barbados. As tobacco production increased in Virginia, the price for tobacco in Barbados fell to three farthings per pound. To quote Hoyos (1978):

> . . . it was at this stage that the Dutch came to the rescue of the Barbadians. They provided the islanders with such sorely needed commodities as salted provisions and clothing at lower rates than the English traders. They supplied cheap loans, inexpensive equipment and insurance at modest rates. They sent the Barbadians on what may be regarded as guided tours to Brazil where they acquired the skills and expertise that were essential for the development of the sugar industry.

Pieter Blower in 1637 was the first man to bring canes to Barbados from Brazil. In the earlier years these canes were used only to produce rum, but from 1642 the Barbadian planters began to manufacture sugar. They found this a difficult art at the beginning but Brazil gave them the knowledge to achieve success. And, when they wanted a work force for the new industry, the Dutch again came to their aid by introducing them to the vast reservoir of labour in West Africa. Many enterprising planters were involved but undoubtedly the outstanding figure in Barbados was Sir James Drax.

Thus, it was through the Dutch traders, not the British colonizers, that the Barbadians commenced commerce with Brazil, the Netherlands, and the slave coast of Africa. (James Drax and his descendants will figure later in this history of sheep in Barbados.)

Could the original wooled sheep have come to Barbados from Brazil? It seems unlikely that any wooled sheep that might have been raised in tropical Brazil would have suffered so from the pasture and climate of Barbados as did the wooled sheep described by Ligon.

Ligon had studied drawing and painting, and he illustrated his book with designs of sugar mills and drawings of plants and trees. In addition, attached to the binding of the book is a 26 cm × 51 cm foldout "Topographical Description" of Barbados (Figure 2.13.2). This provides an outline of the island upon which are inscribed some geographical features, the names and locations of plantations, and drawings of various animals, including two wooled sheep (Figure 2.13.3). The sheep are apparently hornless, white, wooled, free of wool on the face (except for a wool cap between the ears), and are free of wool on the lower legs from a point just above the hocks. The tails are thin, wooled, and extend to about hock level. Their sex is not apparent.

All of the other animals depicted—horses, cows, asses, and dromedary camels—had been introduced to Barbados. Though the feral pigs shown had been hunted to extinction or corralled, a few camels remained. Two sea monsters depicted offshore are, probably, imaginary.

Although Ligon does not say that he drew the animals, the drawing is accepted by historians as "Ligon's map" and is probably the "landscape" that Ligon said that he had drawn (in letters prefacing the first edition). ("Ligon's map" was omitted from the reproduction of the second edition.) The kinds of animals illustrated are those known to have been introduced to Barbados.

Confirmation that the sheep depicted were of the same type that Ligon had described, eaten, and seen sheared would give the drawings credibility as an aid in identifying the wooled ancestor of the Barbados Blackbelly. Ligon's drawing of long-tailed, white-faced, polled, wooled sheep, plus the historical evidence relative to trade routes, would seem to disqualify the

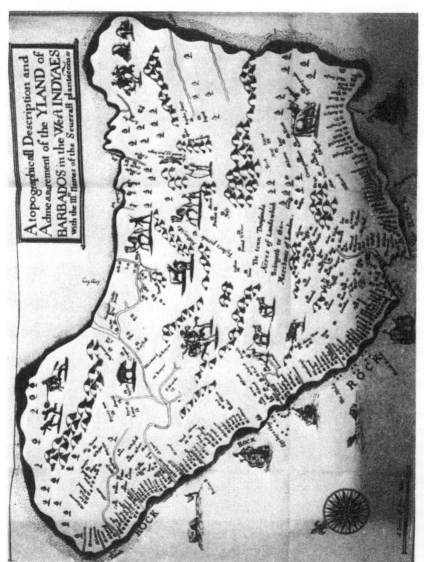

FIGURE 2.13.2 Drawing by Ligon of wooled sheep and other features on map of Barbados in the 1650s.

FIGURE 2.13.3 Close-up of sheep from Ligon's map. Could these be the wooled ancestors of the Barbados Blackbelly?

highly prolific Finnish Landrace, Romanov, and other Northern European short-tail breeds as possible ancestors. The hornlessness of the modern Barbados Blackbelly would tend to disqualify horned breeds. Prolific British breeds do not come readily to mind. Trow-Smith (1957) cited evidence that British sheep in medieval flocks were low in fecundity and concludes that "the marked absence of twinning was almost certainly due to the natural selection of single-lamb ewes."

In his inventory of the estate purchased by Modiford, Ligon mentions cattle, horses, and *assinigoes* (donkeys), but no sheep or other small livestock. Perhaps they were acquired later. In any event, sheep were kept on a neighboring estate, Drax Hall. Ligon was personally acquainted with James Drax, who had been the central figure in the Dutch trade during the decade prior to Ligon's arrival in Barbados. Ligon says that Drax "lived like a Prince" and was one of the few planters who could afford to kill an ox for his table. From prison, and no doubt with senses heightened by nostalgia, Ligon described a great feast at Drax Hall which included "a shoulder of mutton which was there a rare dish."

Thus, it seems certain that Drax kept sheep. Another Drax was to write more than a century later (Lascelles et al. 1786) that sheep were kept "constantly" at Drax's Hope Plantation to produce dung for sugarcane. A traditional local tale is that Drax, of Anglo-Dutch ancestry, personally brought a model of a Dutch windmill from Holland (Dunn 1973). He could have secured wooled sheep from Holland to "improve" the hairy African breed.

Drawings of Dutch sheep breeds (Numan 1835, cited by Bekedam 1974, and reproduced in Bottema and Clason 1979) depict characteristics consistent with Ligon's drawings. The Kemp (Kempische) sheep and the old Flemish (Vlaams) sheep are described as having had long, wooled tails. Kemp sheep are said not to be highly prolific when pastured on heath, but will improve when pastured on grass (Bottema, in conversation, 1979).

Flemish sheep were from the area of Flanders along the border of The Netherlands and Belgium, not far from the area of the Kemp sheep. Flemish sheep are thought to have been of French or Spanish origin. Records of their prolificacy have not been found, but it is reported that when twins were born, one lamb was destroyed. A flock of "Flemish" sheep, based on 20 ewes and 7 rams assembled in 1967 and 1969 from the region where the old Flemish sheep had flourished, averaged 1.94 lambs per litter among one-year-old ewes, 2.56 lambs among two-year-olds, and 2.95 lambs per litter among mature ewes—with virtually 100% conception rates (Bekedam 1974).

The woolless tails (often short), large body size, and prolificacy of several Dutch breeds suggest an intermingling of the highly prolific Northern European short-tail breeds and the larger breeds of Western Europe in The Netherlands. The black and white color pattern of the Dutch Zwartbles is strikingly similar to that of the Romanov and recessive black Finnish Landrace, both exceedingly prolific breeds of the Northern European short-tail group.

Bottema notes (in conversation 1979) that heath sheep were kept in dunging pens to produce dung for fertilizing crops. Also, sheep and goats were tethered individually on small pasture plots, especially by landless workers. This is reminiscent of accounts of dunging pens (Lascelles et al. 1786) recommended for growers of sugarcane in Barbados. Tethering is the predominant practice for sheep in Barbados today.

This circumstantial evidence supports The Netherlands as a source of the prolific, wooled sheep in Barbados described by Ligon.

OTHER SHEEP IN THE AMERICAS

As documented in other chapters of this current work, hair sheep are found in many countries of the Western Hemisphere. The ancestral origins of these sheep may be diverse. Hispaniola (Dominican Republic/Haiti), Cuba, the Virgin Islands, and Jamaica were discovered by Columbus in 1492 through 1494 on three different voyages. His first colony was on Hispaniola, to which he brought livestock (reportedly including sheep) on his second voyage from Spain in 1493. His ships put in at the Canary Islands en route and touched on many of the Caribbean Islands before and after discharging the colonists on Hispaniola.

For 300 years, from the early 16th century to the early 19th century, the importation of slaves from Africa provided opportunities for the introduction of African hair sheep throughout the region. In 1517, Spanish expeditions to Yucatan from Cuba further extended the potential for moving sheep, both European and African.

The English took Jamaica from the Spanish in 1655. By 1664, reports from Jamaica spoke of both tame and wild livestock, including sheep (Sainsbury 1880). By 1670, sheep, goats, and tame hogs were reported in "great plenty" (Sainsbury 1889), and by 1675, "incredible numbers of horses, kine, sheep, goats, etc." were reported in Jamaica (Sainsbury 1893).

Modiford (Governor of Jamaica, previously Ligon's employer in Barbados) wrote in 1666 that an English expedition returning from Nicaragua in 1665 had reported enjoying Spanish beef and mutton and that Indians had told of "many sheep with excellent, fine wool" at the city of "Legovia" near Lake Nicaragua deep in Spanish-held territory (Sainsbury 1880).

When the English captured the Dutch-held island of St. Eustatius in 1665, 500 sheep and goats were found. In 1667, the "main strength" of Dutch Curaçao was reported by the English as being "sheep and goats" (Sainsbury 1880). The Spanish originally settled the island, but the Dutch had held it and the adjacent islands, as well as St. Eustatius, Saba, and part of St. Martins, almost continuously since 1634 and, in the meantime, acquired Surinam on the coast of South America. The Dutch also had occupied the Virgin Islands and had established the first colony on the island of St. Thomas in 1657.

The French were active in the West Indies by the middle of the 17th century. The island of St. Christophers (St. Kitts) had been discovered by Columbus and settled by the English. By 1674, however, St. Kitts had a substantial number of French proprietors who kept 1,687 sheep and 667 cattle, while Frenchmen occupying English properties owned 370 sheep and 268 cattle (Sainsbury 1889). In 1665, a M. le Febvre wrote from Cayenne, French Guiana, to his brother in France (the Abbe' le Febvre), asking for certain supplies "and above all to take care that sheep are sent" (Sainsbury 1893).

Denmark bought St. Croix in 1733 from the Knights of Malta who had acquired it from France in 1653. France had acquired the island in 1651, following its discovery by Columbus in 1493 and subsequent occupations by the Dutch, English, and Spanish. St. Thomas had been discovered by Columbus at the same time and was first settled by the Dutch who later abandoned the area for New Amsterdam (New York). The Danes arrived on St. Thomas in 1666, settled permanently in 1672, and (except for two British interludes between 1801 and 1815) held the island until Denmark sold its Virgin Island possessions to the United States in 1917.

The English were colonizing New England in the early 1600s, and in 1665, it was said (Sainsbury 1880) of Rhode Island and Providence Plantation: "The best English grass and most sheep are in this province, the ground being very fruitful, ewes bring ordinarily two lambs . . . "

What is the origin of these prolific Rhode Island sheep? Rhode Island had been settled in 1636 by religious dissidents from Massachusetts. Thirty-five of the 100 original settlers bound for Massachusetts on the Mayflower in 1620 were of the "Leyden" colony that had initiated the voyage from Leiden (Leyden), Holland, to which they had fled from England nearly a decade earlier. A second contingent from Leiden arrived in 1630. Leiden was a textile center and the men of the colony were largely woolworkers, weavers, and clothmakers (Plooij and Harris 1920). An account of the public execution of five ewes, one mare, one cow, two goats, two calves, one turkey, and a boy of 16 or 17 years in a sensational sexual scandal establishes that sheep and an array of other livestock were in Massachusetts in 1642 (Morison 1952), some five years before Ligon arrived in Barbados.

In May 1623, thirty families of French-speaking Belgians (Walloons) sent out by the Dutch West India Company to establish a Dutch colony arrived in New Netherland (New York). They had sailed by the Canary Islands and then toward "the Wild Coast" (Guiana) to catch the west wind which carried them to their destination (Nicholaes van Wassenaer, quoted in Jameson, 1909). Whether they carried livestock is not stated, but in April 1625, "the worthy Pieter Evertsen Hulft . . . undertook to ship thither (to New Netherland), at his risk, whatever was asked of him, to wit; one hundred and three head of livestock—stallions, mares, bulls and cows—for breeding and multiplying, besides all the hogs and sheep that they thought expedient to send thither; . . . "

In 1627, building on the Leiden connection, the Dutch colony established contact and trade with the English in Massachusetts. In 1630, the Dutch colonists were allowed by the Dutch West India Company to engage in trade along the coast from Florida to Newfoundland. Thus, at a very early date, the Dutch were established on the route of trade with the West Indies and had the demonstrated capability of moving breeding stock safely across the Atlantic Ocean. In 1656, when James Drax and others petitioned England for licenses to import horses and cattle to Barbados, one of the petitioners stated that, formerly, horses had come from Holland and Portugal (Sainsbury 1860).

DEVELOPMENT OF THE BARBADOS BLACKBELLY

Whatever the origin of their wooled ancestors, the sheep of Barbados moved from their precarious beginnings to a place of importance in Barbados agriculture. By 1680, "visitors were delighted by Barbados' . . . fat, sweet mutton fed on sugar cane" (Dunn 1969).

Within 100 years after initial settlement, hair sheep were predominant in the country's sheep production. The Rev. Griffith Hughes wrote (1750):

> The sheep that are natural to this climate and are chiefly bred here, are hairy like Goats. To be covered with Wool, would be as prejudicial to them in these hot climates, as it is useful in Winter Countries for shelter and warmth: Yet as Clothing is necessary (especially in the wet seasons) to the Inhabitants of the warmest Climates, this entire want of Wool upon all Sheep naturally bred here, is abundantly supplied by the Cotton-tree, which yearly, and in great Quantity, produces the finest Wool in the World.

In 1786, Lascelles et al. recommended 48 cattle, 12 horses, and about 40 sheep as adequate to produce sufficient dung to fertilize 40 acres (16 ha) of sugarcane. The authors advised that rams not be kept continuously with the flock. Rams were to be put with the flocks "for a few days" starting the first week of February and were to be replaced every four years. Forty sheep, including 20 ewes, were considered sufficient to maintain the flock and to provide meat "for your own and the table of white servants." Ewes were to be earnotched every January to account for age and were not to be bred after five years of age. Sheep skins were recommended for repairing saddles and horse collars. No mention was made of wool.

The instructions of Henry Drax to Archibald Johnson, estate manager, were appended to the document of Lascelles et al. He was cautioned not to overstock "especially in sheep which are great tainters, and destroyers of pastures, and consequently very injurious to cattle." Fourteen ewes of the "best sort" were enough and ewe lambs "must be constantly killed" and the males castrated and killed "as you have occasion." It may be inferred that the ewes of Barbados were sufficiently prolific to have prompted these cautions against over-expansion of the flock.

Sheep appear to have flourished sufficiently by the middle of the 19th century to sustain great damage by dogs. Schomburgk (1848) reported:

> The breeds of domestic animals degenerate under the tropics. The wool of the sheep becomes in succeeding generations wiry and falls off, and in lieu of a uniform fleece covering, many naked places are observable. [Labourers' dogs] . . . have turned wild, and have increased so rapidly that they commit the greatest ravages in the sheep pens; as many as twenty sheep have been known to be destroyed by these wild dogs in one night.

The description of degenerating fleeces is similar to that which can be observed today in Barbados among sheep of recent crossbreeding with wooled sheep. Buttenshaw (1906) discussed the recurrence of woolliness and speculates on the ancestry of the Barbados "woolless" sheep:

> It should be clearly understood that the present representatives of this breed are far from being purebred. The effect of crossing at some time with imported woolly sheep is constantly to be seen. The offspring of fairly typical woolless sheep are quite likely to develop wool. In fact, it is estimated by an experienced sheep raiser in Barbados, who made a speciality of the woolless sheep, that probably ten percent of the progeny of woolless sheep are more or less woolly, it being quite a common occurrence for one lamb, of two born at a birth, to be quite woolless and the other woolly. From inquiries made into the origin of the wool, it appeared that, though no woolly sheep have been imported in recent years into Barbados for breeding purposes, fifty or sixty years ago American sheep were imported and these no doubt were crossed with hairy sheep already on the island.
>
> Seeing that these hairy sheep are preferred to the woolly kinds by the local butchers, it would seem to be worthwhile for any one who raises them on at all an extensive scale to attempt to breed out the woolliness by careful selection of the best woolless types. This is being done now to a small extent. . . . If this were seriously attempted, however, there would be a distinct danger of in-breeding, since all these animals must be fairly closely related. This could, no doubt, be avoided, and possibly an improvement brought about, by fresh importations from West Africa. I understand that in the Lagos Colony sheep of this kind—that might also be described as "liker goats than sheep"—are very common.

The American sheep introduced 50 to 60 years before Buttenshaw's visit to Barbados could account for the remarks of Schomburgk. Buttenshaw also reported that carcass yields of woolless sheep were superior to those of wooled sheep, and similarly noted the Barbadian butchers' preference for the hair sheep. Buttenshaw commented on the high prolificacy of the woolless sheep.

In the 1930s, Blackhead Persian sheep were introduced to Barbados (Patterson 1976; Stoute, personal communication, 1979). These were fat-tailed hair sheep of low prolificacy and appear to have had little impact on the Blackbelly, although a few black and white sheep, possible descendants of the Blackhead Persian, were observed in Barbados in 1979.

In the 1950s, the Wiltshire Horn was introduced (Patterson 1976). During recent extensive travels in Barbados, the author observed only one horned ewe and some vestigial fleeces; these could be attributed to the Wiltshire Horn which is wooled and horned in both sexes (Mason 1969). It is not uncommon to see spotted, black, white, roan, and tricolor (red, black and white) color patterns among the sheep of Barbados, which could be attributed to the most recent introductions, but also could reflect normal recombination of genes from earlier ancestors.

Suffolk ewes and rams from the United Kingdom and Polled Dorset ewes and rams and Dorset-Cheviot crossbred rams from the United States were introduced for experimental tests in 1978–1979. These sheep exhibited major heat stress and only about 20 to 25% of the ewes conceived during the first year. Respiration rates were about 100 per minute for Blackbelly ewes, 160 for Dorset ewes, and 120 for F_1 ewes, as measured on a few ewes tested at midday in a shed in early September 1978. Fertility appears to have been reduced in both males and females

SELECTION AGAINST WOOLLINESS AND FOR PROLIFICACY

It is the author's opinion that the present-day Barbados Blackbelly combines the hair coat, tasty flesh, extended breeding season, and tropical adaptability of sheep from West Africa with the prolificacy, greater size, and hornlessness of an original wooled stock.

At a conference in 1905 in Trinidad, Buttenshaw (1906) presented a paper and later published the comments of some breeders in attendance, as well as letters from others unable to attend. A breeder from St. Vincent, who maintained a flock of 1,400 sheep, was "opposed to the view of woolless sheep being the best" and insisted that wooled sheep were more profitable for him because of the value of the wool. Another breeder from Grenada recommended a crossbreeding program beginning with wooled rams on "African hairy ewes" to be followed by intermittent "down grading" with "native or hairy" rams to maintain "the necessary balance" of the two breeds. His preference was for Southdown rams, but he also had tried Hampshire, Shropshire, and "Leicestershire."

Clearly, climate was not the only deterrent to the profitable performance of wooled sheep in the Caribbean region: burrs were another such factor. Buttenshaw records that he had learned from breeders in Barbados that hair sheep predominated "by reason of their woolless coats being untroubled by 'burrs' which are so common in most pastures."

A letter, appended by Buttenshaw, from the Hon. F. S. Wigley of the island of St. Kitts made the strongest statement about the burr problem:

> My object when starting the breeding of sheep was three-fold, viz., wool, meat, and manure. For the purpose of the wool I imported on various occasions rams from England and America, but, owing to the prevalence of a grass and a shrub on the estate bearing a burr-like seed rendering the shearing difficult and the wool uncleanable, the breeding for wool had to be abandoned and the breeding of the woolly sheep altogether, they not thriving or fattening readily.
>
> The sheep I now rear are of the "woolless" or hairy breed common to the

West Indies and best adapted to this climate for hardiness, rapid maturity, and weight of carcass; average number 500.

Today, Rosemary Wigley (1978; personal communication 1979) maintains wooled sheep on St. Kitts, but also conducts a weed-removal program following each grazing.

Burrs could have been the kind of ubiquitous environmental factor required to achieve uniformly effective selection against wooliness in the Barbados sheep population. Exposure to burrs probably was intensified increasingly as the culture of sugarcane expanded, forcing sheep off the better land and into the roadsides and onto drier parts of the island and unimproved "rab" lands where burrs abound. The presence of genes for hairiness in the founding gene pool would account for an early, and relatively complete, response to selection against woolliness.

The larger litter size of the Barbados Blackbelly is strong evidence of a prolific ancestor, and almost certainly indicates artificial selection for litter size. The practice of individually tethering almost every sheep on the island creates the opportunity for mating control. In earlier times, and especially before commercial fertilizers, sheep may have been concentrated on various large plantations where castration was recommended (Lascelles et al. 1786) and matings could have been controlled in each plantation breeding unit. The ten coauthors of Lascelles et al. recommended replacing rams only every four years and retaining ewes up to five years, which suggests general acceptance of relatively low replacement rates, with no special recommendations on selection standards.

There could have been conscious selection for larger litters; even the most unsophisticated stockman can count lambs. It also is plausible that single-born lambs were larger, fatter, and more suitable for early slaughter, so that lambs born in multiple births, and thus slower in growth, remained for breeding.

DISTRIBUTION OF SHEEP FROM BARBADOS

From the earliest days of Barbados colonization, reciprocal trade was established with many countries and colonies in the Americas. By 1905 (Buttenshaw 1906), Barbados woolless sheep were "almost entirely raised in Barbados" but "have been distributed by the Imperial Commissioner of Agriculture to most of the other West Indies islands, where they are reported to be doing well and proving useful."

In 1904, the U.S. Department of Agriculture imported four ewes and one ram, all yearlings, to Bethesda, Maryland. A descriptive article by B. M. Rommel of the USDA entitled "Barbados Sheep" and published in the

Breeders Advocate is appended to the paper of Buttenshaw (1906). The article notes that one of the newly arrived sheep grew "a considerable amount" of wool on the shoulders that was attributed to the colder climate of Maryland. However, this finding seems more likely to have been a normal occurrence of woolliness in a breed of mixed ancestry.

There seems to be little evidence to support the claim of Bond (of the U.S. Virgin Islands) that the Barbados Blackbelly "is just a colour variation of our 'native' sheep" (1975, cited by Mason 1980). Mason described the sheep of St. Croix as smaller and less prolific than the Barbados Blackbelly, varying toward white and light colors. There are some parallels between the two breeds, however, as they are reputedly the two most prolific breeds of the West Indian sheep.

The Dutch were the first settlers of the Virgin Islands but they were followed by the Danes, French, English, and Spanish. There is a strong possibility that African and Dutch sheep might have contributed to both the St. Croix and Barbados sheep. Devendra (1972) raised the possibility of Danish ancestry for the St. Croix sheep, but this seems an overstatement.

Patterson (1976) reported that during the 1960s and early 1970s "sizeable numbers" of Barbados Blackbelly sheep were exported from Barbados to Canada, Grenada, Guyana, Jamaica, Mexico, Taiwan, and "other Caribbean countries." The sheep bound for Canada did not arrive due to veterinary complications. After being stopped for some years, exports of Barbados Blackbelly breeding stock have been resumed with shipments to other Caribbean islands and Venezuela.

Recent introductions of European breeds to Barbados have been based on the premise that the Barbados Blackbelly is inferior in carcass merit and that the introduced breeds will improve the carcass characteristics of the Barbados Blackbelly. This premise is questionable. Barbados Blackbelly carcasses are well-muscled, but their long leg bones and almost total lack of carcass fat create the illusion in the live animal of a lack of development of hind quarter as compared to shorter-legged, early-fattening breeds. Visual comparison of imported New Zealand and Australian lamb offered in the retail markets of Barbados reinforced the author's doubts as to the ability of "improved" mutton breeds to enhance the carcass merit of the Barbados Blackbelly. The more likely expectation would be that the improvement would be in the reverse direction. The danger from all mutton-type breeds is that they may reduce the reproductive efficiency of the Barbados Blackbelly, and many of them would increase the fattiness of the carcass under the guise of improving the mutton type.

Local lambs are scarce in Barbados retail markets, as they are consumed mostly by families and neighbors of producers. Only about 2,500 are slaughtered annually by the commercial abattoirs. The author was able to sample the carcass of a Barbados Blackbelly 10-month-old ram lamb, however, and found the meat tender and tasty, with no fatty aftertaste. The flavor brought particular praise from several who previously had been indifferent to the taste of lamb. This tasting experience was consistent with the praise of Ligon (1657); Dunn (1969), who reported comments from 1680; and of present-day Barbadians who are able to get local lamb.

SUMMARY

The evidence reviewed supports the view that the Barbados Blackbelly evolved from crosses of African hair sheep and European wooled breeds. In the case of the Barbados Blackbelly, Ligon (1657) established that the parent stocks necessary for the evolution of a highly prolific, tropically adapted breed had been established in Barbados in the first quarter-century of colonization. The prolific, wooled ancestor of the Barbados Blackbelly has not been identified (although a few wooled breeds are known to be so prolific as to "bear always two Lambs" and to produce an average litter size of two lambs among crosses with an African breed that rarely twins). Evidence is presented that the Dutch had the specially fitted ocean craft, the trading opportunities, and, possibly, the most likely breeds, to have introduced the prolific, wooled ancestors of the Barbados Blackbelly. Speculation that the Barbados Blackbelly might have been introduced "ready-made" from elsewhere is not consistent with available historical or genetic evidence.

REFERENCES

Barnhart, L. (ed.) 1954. *New century cyclopedia of names.* New York: Appleton-Century-Crofts.

Behrens, H., H. Dolhner, R. Scheelje, and R. Wassmuth. 1969. *Lehrbuch der schafzuct.* Hamburg: Verlag Paul Parey.

Bekedam, M. 1974. A crossbreeding experiment with Texel and Flemish sheep; preliminary results. Univ. of Wageningen, Dept. of Animal Breeding. *Proc. Working Symp. Breed Eval. and Crossing Experiments,* Zeist. pp. 407–419.

Bottema, S. en A. T. Clason. 1979. *Het schaap in Nederland.* 174 pp. Zutphen, Netherlands: B.V.W.J. Thieme & Cie.

Buttenshaw, W. R. 1906. Barbados woolless sheep. West Indian Bull., *J. Imp. Dept. Agric. for the W.I.,* 6:187–197.

Devendra, C. 1972. Barbados Blackbelly sheep of the Caribbean. *Tropical Agriculture (Trinidad)* 49(1):23–29.

Devendra, C. 1977. Sheep of the West Indies. *World Review of Animal Production* 13(1):31–38.

Dunn, R. S. 1969. The Barbados census of 1680: Profile of the richest colony in English America. *Wm. and Mary Quart.* 26:3–30.

Dunn, R. S. 1973. *Sugar and slaves—Rise of the planter class in the English West Indies, 1624-1713.* London: Jonathan Cape.

Hoyos, F. A. 1978. *Barbados, a history from the Amerindians to independence.* London: Macmillan Educ. Ltd. (Macmillan Caribbean).

Hughes, Rev. G. 1750. *The natural history of Barbados.* Reprint ed., 1972. New York: Arno Press, Inc.

Jameson, J. F. (ed.). 1909. *Narrative of New Netherland 1609-1664.* New York: Charles Scribner's Sons.

Lascelles, E., J. Colleton, E. Drax, F. Ford, J. Braithwaite, J. Walter, W. T. Holder, J. Holder, P. Gibbs, and J. Barney. 1786. Instructions for the management of a plantation in Barbados and for the treatment of Negroes, etc. London (copy in Barbados Museum).

Ligon, R. 1657. *A true and exact history of the island of Barbados.* (2nd edition, 1673.) London: Peter Parker and Thomas Grey. Reprinted 1976 by Frank Cass, London. 122 pp. (excluding prefaces to 1st ed.).

Mason, I. L. 1969. *A world dictionary of livestock breeds, types and varieties.* 268 pp. Tech. Comm. No. 8 (revised) of the Commonwealth Bur. of Animal Breeding and Genetics, Edinburgh. Commonwealth Agric. Bureaux, Farnham Royal, Bucks., England.

Mason, I. L. 1980. Prolific tropical sheep. FAO Animal Production and Health Paper 17. Rome: FAO.

Maule, J. P. 1977. Barbados Blackbelly sheep. *World Review of Animal Production* 24:19-23.

McDowell, R. E. 1972. *Improvement of livestock in warm climates.* 711 pp. San Francisco: W.H. Freeman and Co.

Morison, S. E. (ed.). 1952. *Of Plymouth Plantation 1620-1647,* by William Bradford. 463 pp. New York: Alfred Knopf.

Patterson, H. C. 1976. The Barbados Blackbelly sheep. 19 pp. Bulletin of Barbados Ministry of Agriculture, Science and Technology, Bridgetown, Barbados.

Patterson, H. C. 1978. The importance of Blackbelly sheep in regional agriculture. Mimeo. 9 pp. Paper presented at Second Regional Conf. on Sheep Prod., Bridgetown, Barbados, Sept. 1978.

Plooij, D., and J. R. Harris. 1920. *Leyden documents relating to the Pilgrim Fathers.* 74 pp. Leyden, The Netherlands: E.J. Brill Ltd.

Sainsbury, N. W. (ed.). 1860. *Calendar of state papers. Colonial series 1574-1660.* Vol. I. London: H.M. Stationery. Krause Reprint, 1964.

Sainsbury, N. W. (ed.). 1880. *Calendar of state papers, Colonial series, America and West Indies, 1661-1668.* Vol. II. London: H.M. Stationery. Krause Reprint, 1964.

Sainsbury, N. W. (ed.). 1889. *Calendar of state papers, Colonial series 1667-1674.* Vol. III. London: H.M. Stationery. Krause Reprint, 1964.

Sainsbury, N. W. (ed.). 1893. *Calendar of state papers, Colonial series 1675-1676 and addenda 1574-1674.* Vol. IV. London: H.M. Stationery. Krause Reprint, 1964.

Schomburgk, R. H. 1848. *The history of Barbados comprising a geographical and statistical description of the island, a sketch of the historical events since the settlement and an account of its geology and natural productions.* Reprinted by Frank Cass, London, 1971. 722 pp.

Stoute, E. A. 1976. The Advocate-News (Barbados) reprint of column on lop-eared, Indian "coolie" goats from shipwreck off Barbados. The *Barbados Times*, 23 January 1866.

Trow-Smith, R. 1957. *A history of British livestock husbandry to 1700*. 286 pp. London: Routledge and Kegan Paul.

Wigley, R. 1978. Experiences of a farmer in the development of a sheep farm started in 1971 in St. Kitts. Mimeo. 6 pp. Paper presented at Second Regional Conf. on Sheep Prod., Bridgetown, Barbados, Sept. 1978.

SECTION THREE:
WESTERN AFRICA

3.1.1 Flock of Forest-Savanna ewes, Ivory Coast

PERFORMANCE OF HAIR SHEEP IN NIGERIA

Almut Dettmers, *University of Ibadan, Nigeria*

INTRODUCTION

Nigeria is the largest of the coastal countries of West Africa. It covers 960,000 km², with 20% of this area in the humid tropics and 80% in the savanna. The humid tropics are characterized by a long rainy season from March to October, with a short dry spell in August. The dry season follows, which is interrupted by rains during December or January. Annual rainfall ranges from 1,000 to 3,000 mm. Temperatures vary little, ranging between 25 and 35° C.

Abundant unimproved pasture, forage, and bush exist in the humid tropical region, but the area is infested with tsetse flies, the carriers of trypanosomes. Therefore, most ruminant livestock (90% of cattle, 80% of sheep and goats) are raised in the north of the country, which is drier and has few tsetse flies.

SHEEP OF NIGERIA

Sheep are an important source of meat in Nigeria; other products (hides, wool, milk) are relatively unimportant. There are about 8 million sheep in Nigeria (Oyenuga 1974), approximately 10 sheep per 100 people; about 18% of all livestock are sheep (Igoche 1974). Average per-head yield has been estimated as 20 kg (Oyenuga 1974), but total yield from 7.5 million sheep was given as 86,000 tons (ASY 1975), implying substantially lower per-head yields.

Although Nigeria is about 80% rural and agricultural, it is not self-sufficient in food. Large quantities of live animals and meat are imported. In 1976, imports included nearly 40 million live animals, 4 million kg of chilled meat, and over 400,000 kg of meat products (Anon. 1977). In the same year, only 187,000 tons of beef and veal, 150,000 tons of sheep and goat meat, 29,000 tons of pork, and 53,000 tons of chicken were produced within the country (ASY 1975).

Considering the shortage of animal protein in the present diet of Nigerians, there is a need to increase livestock production in general and sheep production in particular. Sheep have much in their favor in countries such as Nigeria: their self-sufficiency and utilization of cheap and readily available feeds. They do not require much shelter and thus require only a low investment. Sheep reproduce throughout the year in Nigeria, where day length near the equator does not fluctuate widely.

There are three major breeds of sheep in Nigeria: (1) the Uda and (2) the Yankasa, which are hair sheep in the north (with a small population of Balami in Northeast Nigeria), and (3) the West African Dwarf in the southern humid zone.

West African Dwarf Sheep

A few large flocks of West African Dwarf sheep are maintained on farms of agricultural ministries and universities, but most small flocks are owned by individuals in villages and towns. (See comment on names of West African sheep breeds in Chapter 1.1.)

These sheep are black or black and white spotted, rarely completely white; a few have red coats or red spotting. They are allowed to roam freely; they feed on natural forage and scavenge household scraps and garbage. Matings are rarely controlled.

A recent survey of two villages near Ibadan (Matthewman 1977) reported about 2 to 3 sheep per household, plus 3 to 4 goats and 5 to 8 chickens. The village flocks had a 115% lambing rate (1.15 lambs per ewe lambing), with breeding efficiency estimated at 77% and losses at only 15%. This is a remarkable performance. Observations in these two villages indicated that 90% of the male lambs and about 40% of the young females were sold to local markets at about weaning age; thus nearly two-thirds of the young stock were sold for cash, in addition to about one-fifth of the adult animals. This left only a small proportion of mutton for home consumption—an estimated 5% of total offtake. This meat is not consumed on a regular basis; it is used for ceremonial purposes at births, deaths, and other traditional or religious occasions (Matthewman 1977).

In some institutional flocks, environmental conditions have been slightly improved; however, extensive management continues. Some data have been collected that can serve for an assessment of existing flocks and breeds.

The flock of West African Dwarf sheep at the University of Ibadan has been studied frequently. These sheep have been maintained for nutrition and metabolic studies and for management and carcass studies. Their feeding regime, management, health care, and housing have been described elsewhere (Adeleye 1973; Hill 1960; Dettmers and Loosli 1974; Dettmers et

TABLE 3.1.1. AGE AT FIRST LAMBING OF WEST AFRICAN DWARF EWES

Mean age mo.	Range mo.	Percent
10.0	8-11	11
12.6	12-13	26
15.3	14-17	29
20.6	18-23	17
34.5	24-47	12
58.6	48-72	5

al. 1976a; Igoche 1974). No consistent breeding program has been followed. Ewes breed year-round. Several rams usually run with the flock. Ram lambs are separated from ewes at about three months of age, while ewe lambs remain with ewes and breeding rams. The size of the flock has increased from a few animals in 1950 to about 300 ewes at present. Records have been kept on the ewes' reproductive performance, lamb weights and survival, and growth rate, with live weights recorded at regular intervals. Besides use of animals for research, the main purpose of this flock has been provision of rams for local slaughter. The productivity of the University of Ibadan flock has also been described (Anon. 1950; Hill 1957; Hill 1961); unless otherwise indicated, the following data and comments relate to this flock.

Ewe Productivity. Fully-grown, nonpregnant ewes of the West African Dwarf breed stand about 58 cm high at the withers, are about 55 cm long, and have an average heart girth of 65.5 cm (Dettmers and Loosli 1974). Fully-grown rams have about the same length, but a height of 63 cm and a heart girth of 86.5 cm (Hill 1961).

Breeding efficiency, measured by number of ewes lambing among those exposed, was 77.5% (Dettmers et al. 1976a), a figure similar to the estimate of 77% for village sheep (Matthewman 1977).

West African Dwarf ewes are early maturing. Even though their age at first lambing averaged 20.5 months, with a range from 8 to 72 months, about 37% of them had their first lamb by one year of age—and 66% by 15 months of age (Dettmers and Loosli 1974).

Age distribution at first lambing is shown in Table 3.1.1. These values were for all ewes in 1971-1973. A different group of ewes from the same flock (Orji et al. 1975), which was obviously selected for a particular study, showed less variation.

Table 3.1.2 shows age at first lambing, lambs born per ewe exposed, and lambing interval.

Matings were year-round with lambings averaging three times in two years; lambing intervals ranged from 234 to 248 days (Table 3.1.2). Igoche

TABLE 3.1.2. REPRODUCTIVE TRAITS OF WEST AFRICAN DWARF EWES

x̄	Range	Reference
Age at first lambing, mo.		
–	11 – 14	Hill 1960
14.1	10 – 23	Orji et al. 1975
20.5	8 – 72	Dettmers et al. 1976a
Lambs born/ewes exposed, %		
120.0	–	Hill 1960
140.0	–	Orji et al. 1975
145.6	121 – 200	Dettmers et al. 1976a
Lambing interval, days		
240	–	Hill 1957
234	151 – 571	Orji et al. 1975
248	203 – 277	Dettmers et al. 1976a

TABLE 3.1.3.   LAMBING INTERVAL OF WEST AFRICAN DWARF
                 EWES, DAYS

Interval	x̄
1–2	277
2–3	230
3–4	233
4–5	209
5–6	213
6–7	203
7–8	213

Source: Igoche (1974).

(1974) reported an average interval of 277 days between first and second lambing; as parity advanced and ewes aged, lambing interval decreased to 213 days between the 7th and 8th lambings (Table 3.1.3). Osondu (1969) reported a lambing interval of 304 days, with a range of 74 to 893 days, in a flock at Ado-Ekiti in Western State. He reported an interval of 408 days for the Ibadan flock, with a range of 201 to 665 days. Lambing interval at the Iwo Road Station, a Ministry of Agriculture station near Ibadan, was only 176 days (Ogbonnah 1979). The number of lambs born per ewe was 1.21 at first lambing and increased to nearly 1.9 at fourth lambing (Dettmers et al. 1976a).

Table 3.1.4 shows distribution of prolificacy by parity. Ogbonnah (1979) reported a prolificacy of 132% (1.32 lambs per ewe exposed) for the West African Dwarf flock at Iwo Road Station near Ibadan.

West African Dwarf ewes produced the highest proportion of multiple

TABLE 3.1.4.   PROLIFICACY BY PARITY

Parity	No. lambs	No. ewes	Prolificacy, %
1st	176	145	121.4
2nd	144	98	149.0
3rd	113	71	159.2
4th	84	45	186.7
5th	24	14	171.4
6th	6	4	150.0
7th	6	3	200.0
8th	2	1	200.0

Source: Igoche (1974).

TABLE 3.1.5.   SINGLE AND MULTIPLE BIRTHS IN WEST AFRICAN
DWARF SHEEP, % OF LAMBS BORN

Singles %	Twins %	Triplets %	Reference
76	21	3	Hill 1957
80	20	–	Hill 1960
58	38	4	Orji et al. 1975
37	55	8	Dettmers et al. 1976a

TABLE 3.1.6.   BIRTH WEIGHTS OF WEST AFRICAN DWARF LAMBS
BY SEX AND TYPE OF BIRTH, UNIVERSITY
OF IBADAN, kg

	Hill (1960)		Igoche (1974)		
	No.	$\bar{x}$	No.	$\bar{x}$	SD
Sex					
Male	158	1.81	258	1.74	0.87
Female	147	1.73	297	1.71	1.07
Sex, type					
Male, single	–	2.18	–	1.92	0.90
Male, twin	–	1.45	–	1.63	0.85
Male, triplet	–	–	–	1.52	0.89
Female, single	–	1.72	–	1.93	1.14
Female, twin	–	1.50	–	1.64	0.72
Female, triplet	–	–	–	1.27	0.15

births reported for any Nigerian sheep, with 55% live-born twins and 8% triplets (Dettmers and Loosli 1974; Dettmers et al. 1976a).

Multiple births in the flock appear to have increased over the years (Table 3.1.5). Improved management and other nongenetic factors are the probable causes for these increases, because selection for twinning was not practiced.

Percentage of multiple births increased from about 33% of lambs born to first-parity ewes to 66% for ewes lambing the fourth time. The University of Ibadan flock, with 63% multiple births, equaled the performance of dwarf sheep at Nkwele, Southeast Nigeria (Anon. 1950). This percentage fell within the range of 45 to 64% reported for the flock at Onitsha (Hill 1961). Low values of 21 and 20% reported for twins in the Ibadan flock (Hill 1957; 1960) are similar to values of 20.6% twins and 3.0% triplets in 1961 (Hill 1961) and to 16% twins in 1969 (Osondu 1969).

Additional references on the reproductive ability of the West African Dwarf ewe include Orji and Steinbach (1975; 1976a and b) and Orji (1977).

Lamb Performance. Average birth weights of West African Dwarf lambs rarely exceeded 2 kg (Table 3.1.6). Both single and twin male lambs weighed slightly more than did their female counterparts; male triplets weighed 250 g more at birth than did their female counterparts. There was a slight increase in birth weight with parity of dam (Table 3.1.7). Earlier, Osondu (1969) reported lower birth weights for Ibadan lambs of 1.83 kg for male singles, 1.68 kg for female singles, and 1.44 kg for twins (sex not differentiated). Birth weights at the Iwo Road Station were 1.97 kg for male singles and 1.99 kg for females; male twins weighed 1.78 kg and females only 1.61 kg (Ogbonnah 1979).

Heritability of birth weight for the Ibadan flock was 0.10±0.1 by lamb-ewe regression analysis and 0.35±0.5 by paternal half sib analysis, resulting in a so-called best estimate of 0.11±0.1 (Akinkuolie 1974; Dettmers et al. 1976a).

Survival of lambs is related to birth weight, type of birth, and management of lambs after birth. Hill (1957) reported 6.8% stillborn lambs in the University flock from 1950 to 1954, immediate postnatal deaths of an additional 15%, with an additional 31% dying later. These high losses were attributed to heavy infestation with helminths; more than half of the later deaths occurred within 3 months. Later losses up to weaning for the same flock were 25% (Hill 1960) and 20% (Dettmers and Loosli 1974). Mortality varied with type of birth: 12% for single lambs, 16% for twins, and 46% for triplets (Dettmers et al. 1976a). A flock of West African Dwarf sheep at the Iwo Road Station showed a similar pattern of mortality, with 7.2% dying during the first 5 weeks (three-fourths of these dying during the first week). Losses due to abortion were not included. Twins had a higher death

TABLE 3.1.7.   BIRTH WEIGHT OF LAMBS BY PARITY OF
                    EWES, TYPE OF BIRTH, AND SEX OF LAMB, kg

Parity	Type	Male			Female		
		No.	x̄	SD	No.	x̄	SD
1	S	47	1.71	0.91	60	1.78	1.15
	Tw	32	1.49	0.46	31	1.27	0.86
	Tr	5	1.36	0	1	1.36	0
2	S	21	1.86	1.00	29	2.02	1.13
	Tw	34	1.38	0.18	44	1.49	0.89
	Tr	8	1.36	0.87	8	1.14	0.50
3	S	16	2.41	1.45	13	2.06	0.73
	Tw	30	1.92	1.10	45	1.65	0.87
	Tr	5	1.54	1.02	4	1.48	1.09
4	S	2	2.27	0	7	2.47	1.18
	Tw	36	1.74	0.57	30	1.77	0.59
	Tr	4	2.04	0.50	5	1.18	0.49
5	S	3	1.67	2.61	2	2.04	0.50
	Tw	8	1.76	0.57	8	1.76	0.57
	Tr	2	1.59	0.50	1	1.36	0
6	S	-	-	-	2	2.27	0
	Tw	2	1.59	0.50	2	1.59	0.50
	Tr	-	-	-	-	-	-
7	S	-	-	-	-	-	-
	Tw	2	1.82	0	4	1.71	0.44
	Tr	-	-	-	-	-	-
8	S	-	-	-	-	-	-
	Tw	1	1.82	0	1	1.82	0
	Tr	-	-	-	-	-	-
All	S	89	1.92	0.90	113	1.93	1.14
	Tw	145	1.63	0.85	165	1.64	0.72
	Tr	24	1.52	0.89	19	1.27	0.15

Source: Igoche (1974).

rate than did single lambs (Ogbonnah 1979). Ewe lambs showed heavier
losses than did rams.

Survival and growth of lambs depends largely on milk production by the
dam and on postweaning feed and management. Milk production of ewes in
relation to preweaning growth of lambs was studied by Adu (Adu et al.
1974; Adu 1975): three groups of 6 ewes were respectively supplied with

TABLE 3.1.8.   MILK YIELD DURING 10 WEEKS FOR 3 GROUPS OF
                 WEST AFRICAN DWARF EWES ON 3 DIFFERENT
                 ENERGY-INTAKE LEVELS

| | % of ARC energy standard | | |
Variable	75%	100%	125%
Number of ewes	6	6	6
Milk yield, g/day	321	408	533
Peak yield, g	481	697	670
Total 10 week yield, kg	159	238	251

Source: Adu et al. (1974); Adu (1975).

TABLE 3.1.9.   LIVE WEIGHTS OF WEST AFRICAN DWARF SHEEP, kg

Age, mo.	No.[a]	Weight	No.[b]	Weight
1	28	5.0	–	–
2	24	6.6	205	5.7
3	29	8.7	178	8.2
6	105	11.3	139	11.7
9	94	14.7	117	14.6
12	81	16.8	76	16.8
15	79	20.4	70	18.2
18	48	23.3	63	19.8
24	45	28.4	47	24.1
36	53	27.2	22	25.4
48	30	30.3	–	–
60	24	30.1	–	–
72	12	30.8	–	–

[a] Akinkuolie (1974); females only.
[b] Hill (1960); males and females.

75%, 100%, and 125% of Agricultural Research Council (1965) energy standards for sheep during the latter part of pregnancy and during lactation. Milk yields of the ewes differed substantially (Table 3.1.8), as did birth weight and gains of their lambs.

Growth. Table 3.1.9 shows the live weights of lambs and their daily gain before and after weaning in the Ibadan flock during different periods (Hill 1960; Akinkuolie 1974). The more recent improvement in gains of the sheep in the University flock is probably due to management and nutrition.

Carcass Traits. Some information is available on carcass traits of the West African Dwarf sheep (Hill 1960; Taiwo 1974; Nkemeatu 1974; Dett-

TABLE 3.1.10. RETAIL CUTS FOR TWO GROUPS OF WEST AFRICAN DWARF EWES, kg

Variable	Group 1 [a]			Group 2 [b]	
	$\bar{x}$	SD	%	$\bar{x}$	%
Age, months	43.2	22.1		35.4	
Yield, %	42.5	4.1		44.0	
Leg	3.0	0.11	35.7	3.0	32.2
Shoulder	1.7	0.40	20.3	1.8	19.6
Rack	1.4	0.32	16.7	1.3	14.1
Loin + flanks	0.9	0.21	10.7	1.0	10.3
Shanks + breast	0.8	0.21	9.5	0.9	9.9
Neck	0.6	0.12	7.2	0.6	6.5

[a] Dettmers et al. (1976b); [b] Nkemeatu (1974).

mers and Loosli 1974; Adeleye and Oguntona 1975; Dettmers et al. 1976b). Dressing percentages ranged from 40 to 45% for sheep from 1 to 4 years of age and older. Carcass weights in these studies ranged from 4.2 to 9.9 kg, but most were from 6.2 to 8.2 kg. These reports also contain information on organ weights and weights of various other parts (blood, hide, etc.).

Nkemeatu (1974) reported results on 45 sheep slaughtered at ages from 1 to over 4 years. Mean live weights for the different age groups ranged from 12.4 to 21.5 kg. Dressing percentage changed little with age (averaging 43.4%), with carcass weight increasing from 4.8 kg for the youngest group to 8.8 kg for the oldest group.

Retail cuts and their percentages of cold carcass for 30 ewes between 3 and 4 years old (Dettmers et al. 1976b) are shown above (Table 3.1.10) with similar information for 7 additional ewes from the same flock (Nkemeatu 1974).

Young ewes had superior carcasses compared to rams of the same age (Table 3.1.11). The four primal cuts (shoulder, rack, loin, and leg) made up about 85% of the carcass. Carcasses for this group of West African Dwarf sheep included 51.3% hind saddle (leg, loin, flank) and 48.9% fore saddle (rack, shoulder, shanks, breast), without the neck. However, the neck is considered a delicacy, thus may be included among the primal cuts. Young ewes exceeded rams of the same age in yield (44.1% to 39.9%) and in red meat. Primal cuts from ewes contained 67% muscle, 7.3% fat, and 25.7% bone; values for rams were 65.3% muscle, 6.6% fat, and 28% bone (Dettmers et al. 1976b). Retail cuts were analyzed for chemical composition (Nkemeatu 1974) as were organs (Adeleye and Oguntona 1975).

Crossbreeding with Permer Sheep. In 1973, a small flock of 6 rams and 20 ewes of Permer sheep were donated to the University of Ibadan. Permer

TABLE 3.1.11.   CARCASS COMPOSITION OF YOUNG WEST AFRICAN
                  DWARF SHEEP

Variable	Ewes			Rams		
Number	8			13		
Age, months	13			15		
Live weight, kg	13.3			15.7		
Cold weight, kg	6.5			6.2		
Yield, %	44.1			39.9		
Composition of cuts, %	M[a]	B	F	M	B	F
Leg	73.3	20.2	6.5	71.7	22.8	5.5
Shoulder	69.1	22.1	8.8	68.5	23.8	7.7
Rack	65.8	30.5	3.7	57.9	35.4	6.7
Loin + flanks	68.7	19.1	12.2	67.5	24.3	8.2
Shanks + breast	58.2	31.4	10.4	61.7	31.7	6.6
Neck	66.8	31.1	2.1	64.5	30.8	4.7

[a] M = Muscle, B = Bone, F = Fat.

were developed from a cross of Blackhead *Per*sian fat-rump sheep with
Mutton *Meri*no and *Meri*no Landrace from West Germany. These crossbreds were intended for use in Sudan as foundation stock for the development of hair sheep with improved mutton production in the arid zones of
Africa (Haring and Mukhtar 1971). The Permer were a genetically segre-
gating flock showing the pronounced blackhead pattern, with completely
white or spotted coats. Some sheep had fat tails and others none at all; some
had a hair coat, while others produced wool that would later be shed. Inter-
estingly, these sheep which had been breeding seasonally in Germany
adapted to a nonseasonal breeding cycle within a year of introduction to
Nigeria (Dettmers 1973).

     Some of the Permer sheep at Ibadan were bred *inter se* to study their
adaptation to Nigerian conditions; two rams also were mated to ten previously parous West African Dwarf ewes. There were no mating or lambing
difficulties, even though Permer rams weighed about 70 kg and were twice
as heavy as the West African ewes. Of the 20 ewes exposed, 14 lambed
(70%) and 23 lambs were born (115%). The ewes that lambed had a prolificacy of 164%. Of the lambs born, 26% were single, 61% twins, and 13%
triplets. Birth weights of crossbred lambs did not differ due to sex of lamb or
type of birth (Ukaegbu 1976). Birth weights of crossbreds (2.3 kg) not only
corresponded to the average of parental breeds but exceeded the weight of
the smaller parent breed (1.6 kg) by more than 40%. The superiority in
weight over the smaller breed decreased to 5% at the age of 1 month, after
which crossbreds generally weighed less than did West African Dwarf

TABLE 3.1.12.   COMPARISON OF SOME REPRODUCTIVE TRAITS OF THE LOCAL FOREST TYPE SHEEP WITH THE NUNGUA BLACKHEAD OF GHANA

Trait, %	Forest Type	Nungua Blackhead
Fertility	79.3	72.0
Prolificacy	151.0	102.3
Lambs born alive	99.0	97.0
Lambs weaned	84.0	94.3
Ewe productivity	118.7	85.7

Source: Ngere (1973).

lambs. More than 50% of the crossbred lambs did not survive (Dettmers et al. 1976c); all lambs that died were twins and triplets. The 8 triplets that reached 1 year of age weighed an average of 18.6 kg, as compared to 31.7 kg for the Permer and 16.8 kg for West African Dwarf (Dettmers 1974; Dettmers et al. 1976c).

A new crossbreeding experiment has begun under well-controlled and standardized conditions, in which West African Dwarf ewes are being crossed with Permer, Uda, and Yankasa rams. The crossbred lambs are to be compared with purebred lambs of the same rams in the same environment (Taiwo 1978).

Other Dwarf Sheep in West Africa

In Ghana, dwarf sheep are called "forest sheep." They resemble the Nigerian variety in coat color and size. Table 3.1.12 lists some reproductive traits for sheep from the flock at the University of Ghana Experiment Station, Nungua (Ngere 1973). Performance traits for the flock are defined as follows: fertility, number of ewes lambing as a percent of ewes exposed; prolificacy, number of lambs born as a percent of ewes lambing; lambs born alive as a percent of lambs born; lambs weaned as a percent of lambs born; ewe productivity, number of lambs weaned as a percent of ewes exposed.

The Nungua sheep are indigenous sheep of Ghana upgraded to Blackhead Persian. Crossbreeding was not beneficial, except for lamb survival to weaning. For a flock of 47 forest-type ewes the rate of multiple births was 77.7%: 71.1% twins and 6.6% triplets. Of these 47 ewes, 45 lambed (95.7%). These findings contrast sharply with those for the Nungua Blackhead flock of 99 ewes, which had a lambing rate of 66% and only 15% twinning. The Nungua had an advantage in growth rate and carcass weight, as would be expected from a breed of larger mature size. Ngere (1973) reported live and carcass weights of 12.9 and 5.2 kg for 10 forest-type sheep; corresponding weights were 20.0 and 8.4 kg for 22 Nungua sheep.

In the Cameroons, the small forest sheep are called Djallonké (Doutressoulle 1947); they originated from the Fouta-Djallon of the West African coast. Coastal varieties have live weights of 25 kg, whereas plateau sheep living at 1,000 to 2,000 m altitude weigh about 30 kg (Vallerand and Branckaert 1975). The forest sheep stands about 67 cm high. Its color is predominantly combinations of black and white; rarely is it entirely white, red, or red-spotted. Rams usually are horned.

A study of 140 ewes at the Nkolbisson Station, about 10 km from Yaoundé, reported fertility of 96%, and only one of the ewes was sterile. Most of the ewes lambed three times within two years. No triplets were born in this flock. Twinning was recorded only for "two outstanding sisters," which together produced 9 pairs of twins after each had had 8 lambings.

At the Nkolbisson Station in the Cameroons, daily gains of 64 to 85 g for males and of 57 to 79 g for females were obtained over a feeding trial period from 30 to 150 days of age, by supplementing lambs with three different concentrates (Vallerand and Branckaert 1975). Lower gains of 52 g for males and 45 g for females were obtained under extensive management, without supplementation.

Mortality of lambs before one year of age for all lambs combined was 32.2% for the period 1966–1970 and 24.1% for 1971–1974. Twins died at a rate of 44% and 33%, respectively, and singles had a mortality of 29% and 21% for the two periods at the Nkolbisson Station (Vallerand and Branckaert 1975). These mortalities exceeded those reported for dwarf sheep in Nigeria. During the first 30 days of lactation (under extensive management) in the earlier years (1965–1969), ewes gave between 85 and 115 g of milk per day. More recently (1970–1973) they gave between 112 and 134 g, probably due to improvement in management. When the sheep were well supplemented, lactation occurred for 117 (±6) days with an average total yield of 87 kg milk.

From a study of 23 rams slaughtered at different live weights, a 40% dressed yield was obtained from rams with a live weight of 18 kg, 43% at 20 kg, 47% at 22 kg, and 48.5% at 28 kg slaughter weight. Age at slaughter was not stated.

SHEEP OF THE ARID ZONE (UDA AND YANKASA)

In Nigeria, there are two distinct Northern breeds of sheep, the Uda (Ouda) and the Yankasa. They are long-legged hair sheep with strong legs, and nomads herd them either with cattle or in separate herds. The front half of the Uda's body is black and the hind part is white. The Yankasa (also called the White Fulani sheep) is predominantly white with black spotting around

TABLE 3.1.13. EWE PERFORMANCE OF PUREBRED YANKASA (Y) AND UDA (U) SHEEP AND THEIR CROSSES WITH MERINO (M) AT THE SHIKA STATION (1959-61)

Breed or cross	Y	U	M×Y	M×U
Number ewes	70	56	113	47
Lambings	102	71	147	64
Singles	76	61	121	55
Twins	26	10	26	9
Litter size	1.25	1.14	1.18	1.14
Mortality, %	3.9	7.4	1.2	0
Lambing interval, days	236	270	284	273

Source: Ferguson (1964).

TABLE 3.1.14. BODY WEIGHT OF PUREBRED YANKASA (Y) AND UDA (U) SHEEP AND THEIR CROSSES WITH MERINO (M) AT SHIKA, kg

Breed or cross	Y		U		M×Y		M×U	
	M	F	M	F	M	F	M	F
No.	37	24	34	30	40	31	14	32
Birth wt	3.5	4.0	3.9	3.6	3.9	3.7	4.4	4.0
No.	24	24	20	20	36	31	19	31
Wt, 3 mo	14.9	13.9	16.2	15.3	18.5	14.3	18.5	17.2
No.	10	23	5	7	32	28	12	20
Wt, 12 mo	29.8	21.2	32.6	27.1	34.3	27.0	33.6	29.8
No.	3	10	2	2	17	23	5	20
Wt, 18 mo	46.9	30.0	36.1	24.3	46.3	33.3	46.9	34.8
No.	–	–	–	1	7	14	2	12
Wt, 24 mo	–	–	–	30.0	54.7	35.5	54.9	40.1

Source: Ferguson (1964).

the eyes. Their birth weights are 3.5 kg. Yearling weights of 25 to 29 kg contrast with the 17 kg reported for the West African Dwarf (Dettmers and Hill 1974).

Table 3.1.13 shows the reproductive performance of both breeds (U and Y) for three years at the Shika Station, Kaduna State, Northern Nigeria, along with comparable data for F_1 crosses of indigenous sheep with Merino rams (Ferguson 1964). The same report by Ferguson contains information on weights of progeny from these matings at ages up to 24 months (Table 3.1.14).

TABLE 3.1.15.   REPRODUCTIVE PERFORMANCE OF INDIGENOUS
                 UDA AND YANKASA EWES, THEIR BACKCROSSES
                 TO SUFFOLK, AND HALF-BREED INDIGENOUS
                 CROSSES AT THE SHIKA STATION, 1977

Breed or cross	U	SSU[a]	Y	SSY[a]	C[b]
Ewes exposed	28	26	39	39	23
Ewes lambed	25	26	31	34	18
Lambs born	36	31	39	41	22
Lambing, %	129	119	100	105	96
Weaning, %	71	89	72	62	91
Mortality, %	44	26	28	41	5
Litter size	1.44	1.19	1.26	1.21	1.22
Birth weight, kg	3.4	4.1	3.0	3.1	3.4
Daily gain, g	89	108	85	105	126

[a] SSU and SSY from Suffolk-Yankasa and Suffolk-Uda ewes
     backcrossed to Suffolk rams.

[b] Suffolk F_1 and F_2 of both kinds combined and Suffolk x Uda or
     Yankasa half-breeds.

Source: Adu and Brinckman (1978).

Dressing percentages for a total of 7 rams from these groups slaughtered
at 13 to 14 months of age ranged from 38 to 41%, yielding carcasses of up to
11.5 kg. (The slaughtered rams weighed less than those kept, based on the
12-month weights in Table 3.1.14).

Table 3.1.14 indicates that the Yankasa and Uda breeds are larger and
grow more rapidly than do the West African Dwarf sheep. Birth weights are
approximately double those of the West African Dwarf. Crossing with the
Merino increased growth rate and size, particularly for weights of ewes at
12 and 18 months.

Lambing percentage is defined as 100 × (lambs born/ewes bred); the
percentage for weaning or ewe productivity is 100 × (lambs weaned/ewes
exposed). Mortality percentage is 100 × (lambs died [from birth to weaning
at 4 months]/lambs born). The twinning rate is included in the lambing percentage. No triplets were born. First crosses (C) were clearly superior in
terms of percentage of lambs weaned, reduced mortality, and higher
average daily gain (Table 3.1.15).

SUMMARY

Indigenous sheep in the humid tropics of Nigeria are called West African
Dwarf; they are similar to the forest or Djallonké sheep in other West

TABLE 3.1.16.   REPRODUCTIVE PERFORMANCE OF NIGERIAN
INDIGENOUS SHEEP

Breed	Location	Lambing %	Weaning % 4 mo	Mortality % preweaning	Twinning %
Uda	Shika[a]	129	71	14	30
Yankasa	Shika[a]	110	78	12	12
Uda	Katsina[b]	107.9	68	16	23.4
Yankasa	Katsina[b]	110	70	14	22.2
Balami	Marguba[a]	142.5			
West African Dwarf	Ibadan[c]	121.4	70	20	20.8

[a] Adu (unpublished report).
[b] Mani and Adu (unpublished report).
[c] Ngere (unpublished report).

African countries. Ewes are nonseasonal and can produce three crops in two years. They are early maturing, show good fertility, and have high prolificacy (multiple births). However, in spite of their adaptation, mortality rates are high. Most sheep are maintained under extensive management. Supplementation has increased growth rates, even though these sheep are small with one-year weights of about 17 kg. These sheep yield fair carcasses with over 40% yield.

Among sheep of the arid zone, the Uda and Yankasa of Northern Nigeria are described. The Yankasa is generally superior to the Uda and to the West African Dwarf Sheep, especially in reproductive performance.

ADDENDUM

After this report was written, the following summaries from surveys (Ngere et al. 1979) were made available, including performance data for different breeds and crosses throughout Nigeria. These data are presented in Tables 3.1.16 and 3.1.17.

TABLE 3.1.17. GROWTH AND CARCASS CHARACTERISTICS OF NIGERIAN INDIGENOUS SHEEP

Breed	Birth wt, kg		Daily gain g						Dressing %	Fat[b] g	Prime cuts %
			0-3 months		0-6 months		0-1 year				
	S[a]	T[a]	S	T	S	T	S	T			
Uda	3.8	3.4	156	111	109	95	98	91	43-49	296-597	65.5
Yankasa	3.2	2.9	136	103	100	83	94	80	40-51	400-759	65.6
Uda	3.3	3.0	138	112	91	78	81	68			
Yankasa	3.6	3.0	148	116	99	82	80	72			
Balami	3.5	3.1	131	120	110	99	100	93			
WAD	1.7	1.5	84	77	64	50	65	29	43.6		85
WAD	2.5	2.1	100	77							

a S = singles, T = twins.
b Fat = Kidney + Pelvic fat.

REFERENCES

Adeleye, I.O.A. 1973. Sheep production research information brochure. Department of Animal Science, University of Ibadan.

Adeleye, I.O.A. and E. Oguntona. 1975. Effects of age and sex on live weight and body composition of West African Dwarf sheep. *Niger. J. Anim. Prod.* 2:264–269.

Adu, I. F. 1975. The effects of "steaming up" on the birth weight, lactation, and growth of West African Dwarf sheep. Ph.D. thesis. University of Ibadan.

Adu, I. F. and W. L. Brinckman. 1978. Crossbreeding of local ewes. *Ann. Rept. of Research in Anim. Prod.*, pp. 24–25. NAPRI, Shika.

Adu, I. F., E. A. Olaloku, and V. A. Oyenuga. 1974. Effects of energy intake during late pregnancy on lamb birth weights and lactation of Nigerian Dwarf sheep. *Niger J. Anim. Prod.* 1:151–161.

African Statistical Yearbook (ASY). 1975. New York: United Nations.

Agricultural Research Council (ARC). 1965. The nutrient requirements of farm livestock, No. 2: Ruminants. London: Agricultural Research Council.

Akinkuolie, K. 1974. Heritability of birth weight and the growth pattern of West African Dwarf sheep. Student Research Report, Department of Animal Science, University of Ibadan.

Anon., 1950. Nigerian sheep and goat production. Division of Animal Husbandry and Health, Faculty of Agric., University College, Ibadan.

Anon., 1977. Nigeria trade summary 156–76, Fed. Government, Lagos.

Dettmers, A. 1973. Report on the Permer sheep project. Mimeo. Department of Animal Science, University of Ibadan.

Dettmers, A. 1974. Report on the Permer Sheep Project. Mimeo. Department of Animal Science, University of Ibadan.

Dettmers, A. and D. H. Hill. 1974. Animal breeding in Nigeria. *Proc. First Wld. Cong. Genet. Applied to Livestock Production* 3:811–820. Madrid.

Dettmers, A. and J. K. Loosli. 1974. Live performance and carcass characteristics of West African Dwarf sheep. *Niger. J. Anim. Prod.* 1:108. Abstract.

Dettmers, A., C. A. Igoche, and K. Akinkuolie. 1976a. The West African Dwarf sheep. I. Reproductive performance and growth. *Niger. J. Anim. Prod.* 3(1):139–147.

Dettmers, A., J. K. Loosli, B.B.A. Taiwo, and J. A. Nkemeatu. 1976b. The West African Dwarf sheep. II. Carcass traits and mutton quality. *Niger. J. Anim. Prod.* 3(2):25–33.

Dettmers, A., O. Oduwole, and T. A. Anwana. 1976c. Heterosis in animal production. *Niger. J. Anim. Prod.* 3:91. Abstract.

Doutressoulle. 1947. *L'élevage en Afrique occidentale française.* Paris: Maisonneuve et Larose. Cited by Vallerand and Branckaert, 1975.

Ferguson, W. 1964. The development of sheep and goat production in the Northern region of Nigeria. First FAO Afr. Regional Meetg. Anim. Production and Health, Addis Ababa.

Haring, F. and A.M.S. Mukhtar. 1971. Kreuzung afrikanischer haarschafe mit deutschen Merinos für Ostafrika. *Z.f. Säugetierkunde* 36:304–315.

Hill, D. H. 1957. Quinquennial report (June 1950–June 1955), Sheep Section, Division of Animal Husbandry and Health, Faculty of Agric., University College, Ibadan.

Hill, D. H. 1960. West African Dwarf sheep. Third Ann. Conf. Sci. Assoc. Nig., Ibadan.

Hill, D. H. 1961. Nigerian Dwarf sheep. Fac. Agric., University of Ibadan.

Igoche, C. A. 1974. Lambing performance of the West African Dwarf sheep. Student Research Report, Department of Animal Science, University of Ibadan.

Matthewman, R. W. 1977. Small livestock production in two villages in the forest and Derived Savanna zones of Southwestern Nigeria. *AES Res. Bull. No. 1,* Department of Agric. Extension Services, University of Ibadan.

Ngere, L. O. 1973. Size and growth rate of West African Dwarf sheep and a breed, the Nungua Blackhead of Ghana. *Ghana J. Agr.* 6:113–117.

Ngere, L. O., I. F. Adu, and I. Mani. 1979. Report of Subcommittee on Small Ruminants; Ad-hoc Committee on National Livestock Breeding Policy. Pp. 13–18. NAPRI, Shika.

Nkemeatu, J. A. 1974. Carcass composition of dwarf sheep and goats. Student Research Report, Department of Animal Science, University of Ibadan.

Ogbonnah, R. O. 1979. The birth weight of West African Dwarf lambs at Iwo Road Dairy Farm. Student Research Report, Department of Animal Science, University of Ibadan.

Orji, B. I. 1977. Studies on the reproductive biology of the Nigerian Dwarf sheep. Ph.D. thesis, University of Ibadan.

Orji, B. I. and J. Steinbach. 1975. Studies on the biology of reproduction of the Nigerian Dwarf sheep. *Niger. J. Anim. Prod.* 2:68. Abstract.

Orji, B. I. and J. Steinbach. 1976a. Puberty in ewe lambs of Nigerian Dwarf sheep. *Niger. J. Anim. Prod.* 3:97. Abstract.

Orji, B. I. and J. Steinbach. 1976b. Lamb nursing in the Nigerian Dwarf sheep. *Niger. J. Anim. Prod.* 3:98. Abstract.

Orji, B. I., J. Steinbach, E. A. Olaloku, and I.O.A. Adeleye, 1975. The West African Dwarf sheep. I. Reproductive performance of the Nigerian dwarf sheep. *Bull. Sci. Assoc. Nig.* 1:35. Abstract.

Osondu, J. N. 1969. Production characteristics of the Nigerian Dwarf sheep in the western and midwestern states of Nigeria. Student Project, University of Ibadan.

Oyenuga, V. A. 1974. Feeds and foods in tropical Africa. Pp. 27–42 in *Animal Production in the Tropics.* Loosli, J. K., V. A. Oyenuga, and G. M. Babatunde, eds. Ibadan: Heinemann Educational Books (Nigeria) Ltd.

Taiwo, B.B.A. 1974. Carcass traits in dwarf sheep. Student Research Report, Department of Animal Science, University of Ibadan.

Taiwo, B.B.A. 1978. The performance of the West African Dwarf sheep and its improvement by crossbreeding. Unpublished.

Ukaegbu, O. S. 1976. Heterosis in crossbred sheep? Student Research Report, Department of Animal Science, University of Ibadan.

Vallerand, F. and R. Branckaert. 1975. La race ovine Djallonké au Cameroun. Potentialités, zootechniques, conditions d'élevage, avenir. *Rev. Elev. Med. Vét. Pays Trop.* 28:523–545.

REVIEW OF HAIR SHEEP STUDIES IN SOUTHWESTERN NIGERIA

A. A. Ademosun, K. Benyi,
O. Chiboka, and C. M. Munyabuntu,
University of Ife, Nigeria

Nigeria is divided into five grassland zones (from north to south): (1) the Sahel, (2) Sudan Savanna, (3) Northern Guinea Savanna, (4) Southern Guinea Savanna, and (5) Derived Savanna.

Southwestern Nigeria, as discussed here, occupies part of the Southern Guinea Savanna and part of the Derived Savanna. The Southern Guinea Savanna (also called the Low-Tree Savanna) is a transitional zone between the tropical rain forest and the savanna belts, with low trees forming the climax vegetation. Rainfall ranges from about 1,150 mm to 1,500 mm between March and November (for a period of about 100 to 120 days in the southern part, and for up to about 150 days in the northern part) (Adeniyi 1974).

Common grass species in the region include *Pennisetum purpureum, Andropogon tectorum,* and *Panicum maximum.* The Derived Savanna is open parkland intermediate between the rain forest belt and the Southern Guinea Savanna. The vegetation is made up of tall grasses, primarily *Pennisetum, Andropogon, Panicum, Chloris, Hyparrhenia, Paspalum, Cynodon, Melinis,* and herbaceous plants (Olubajo 1974). Legumes include *Calopogonium mucinoides* and *Pueraria phaseoloides* (Ademosun 1973). The area has high temperatures and humidity and is tsetse fly infested. Common crops include yams, cassava, and grains.

HAIR SHEEP IN NIGERIA

Of the 8 million sheep in Nigeria, about 30% are of the West African Dwarf sheep breed, which is the predominant breed in the parts of southwest Nigeria with over 1,000 mm annual rainfall. This is partly because these sheep are more tolerant to trypanosomiasis than are the other sheep breeds in Nigeria.

Sheep meat is well accepted in Nigeria (Loosli and van Blake 1974); sheep and goats, combined, supply about 30% of the meat consumed. The skin of the West African Dwarf breed is of poor quality and is usually eaten instead of tanned. Mature live weights of the West African Dwarf sheep average 20 to 25 kg at 2 years of age; the sheep scavenge and fend for themselves within human settlements (Dettmers et al. 1976).

Traditionally, sheep have been viewed as slow-growing scavengers and poor converters of feed. As a result, they have been neglected in livestock development planning.

FEEDING AND MANAGEMENT

Pasture has been accepted as the cheapest source of nutrients for ruminants. However, herbage is expensive to produce in Southwestern Nigeria; its nutritive value and productivity fall rapidly with maturation. Methods of conservation are poor and pasture production competes with food crop production. In other ways, too, people and livestock compete for food crops.

Apart from government farms and research centers, commercial farms, and teaching farms, there is no organized system of feeding and management. A sheep unit was established at the University of Ife farm in 1968 with stock purchased from local markets. This farm has an intensive system of management involving all-day housing, except when the sheep are put on pasture in adjacent paddocks during pen-cleaning times (from 0800 to 1200). Depending on the ages of the animals, they are fed supplements of 0.5 to 0.75 kg of concentrates per animal per day. Water and salt lick are available ad libitum. The animals are dewormed every 3 months using phenothiazine (Ademosun 1973). Table 3.2.1 shows the mean live weights for different age groups of the animals for 1973 (approximately 100 head) and 1979 (204 head).

In village flocks of the area, lack of an organized system of feeding and management leads to slow growth rates and can contribute to weight loss during the dry season. The practice of allowing the young to roam with the dams contributes to high lamb mortality, especially during the rainy season. Because the animals are susceptible to pneumonia, fencing is needed so that they can be driven into barns for shelter when it rains.

Ewes nurse their lambs even after the ewes have been bred; thus lambs may be forced to stop suckling because no more milk is being produced. At the University of Ife, weaning of lambs between 8 and 13 weeks of age has reduced lamb mortality.

Nutritional investigations have shown that intake of the local forages is

TABLE 3.2.1.   WEIGHT OF WEST AFRICAN DWARF SHEEP AT THE UNIVERSITY OF IFE FARM, kg

	1973							
	Age in weeks							
Class	0	4	8	10	12	16		
All	2.4	5.8	8.6	9.7	10.5	12.1		
Males	2.5	5.9	8.7	9.8	10.9	12.5		
Females	2.3	5.6	8.4	9.4	10.2	11.8		
Singles	2.5	6.6	9.5	10.7	11.9	13.3		
Twins	2.1	5.2	7.3	8.1	9.1	11.2		

	1979[a]							
	Age in months							
Class	0	4	8	10	12	16	20	24
All	na[b]	9.5	13.9	19.2	23.3	24.2	30.1	33.1
Males	na	11.1	14.2	24.3	24.7	26.3	36.5	39.1
Females	na	8.0	13.9	14.0	21.8	22.0	23.7	27.1

[a]   Weights for 1979 available on the basis of age in months.
[b]   Weights for singles and twins not available for 1979.
      na = not available.

low. A number of workers have shown that forage intake by sheep in this region is much higher than that by goats.

Ademosun (1970a) observed that sheep on *Pennisetum purpureum* tended to digest that forage better than did goats ($P < 0.05$) and also showed higher ($P < 0.01$) intake of the forage. However, goats digested the various components of *Stylosanthes gracilis* better than did sheep, although, in terms of nutritive value index (NVI), this difference in digestibility was offset by the greater feed intake by sheep (Ademosun 1970b). The mean NVI shown by goats on a stylo forage was 39.8, while that by sheep was 54.6 (Ademosun 1970b). The maintenance requirement was estimated to be about 1.38 kg of total digestible nutrients (TDN) per 100 kg body weight for goats and 1.86 kg for sheep (Ademosun 1970b).

Sheep and goats are often thought to have similar nutrient requirements; the above work, however, indicates that sheep have higher TDN requirements.

At the University of Ife, Munyabuntu (1977) reported that sheep consumed much more ($P < 0.01$) *Cynodon* forage than did goats. Generally, the digestibility rates were similar. The higher intake level for sheep also led to higher NVI values for sheep than those for goats ($P < 0.01$).

At the University of Ibadan, Adegbola et al. (1976) established that hair sheep in this region have a digestible crude protein (CP) requirement for maintenance ranging from 0.41 to 1.20 g per day per kg metabolic body weight. The same study showed that endogenous losses of nitrogen and nitrogen requirements were low compared with losses of other breeds of sheep; this was attributed to the adaptation of West African Dwarf sheep to low levels of nutrition.

Adu et al. (1974) found that varying energy intake for ewes during late pregnancy affected daily gains during gestation and lactation. Pregnant ewes weighing 15 to 20 kg were fed rations A, B, and C containing 75, 100 and 125% of the Agricultural Research Council (ARC 1965) energy standards for sheep for late pregnancy and lactation. During the last seven weeks of pregnancy dry matter (DM) intake was 569 g per animal per day for ration A, 748 g for ration B, and 826 g for ration C. This was calculated to be 0.99, 1.81, and 2.16 Mcal of metabolizable energy (ME) per day for rations A, B, and C, respectively. During lactation, the intakes for rations A, B, and C were 660, 748, and 857 g of DM; 287, 593, and 690 g of digestible organic matter, and 1.05, 2.16 and 2.51 Mcal of ME.

Higher energy intake increased milk production and ewe weights. There was a correlation of 0.83 between ewe weight prior to lambing and lamb birth weight. The respective gains for rations A, B, and C were 14, 99, and 90 g per day per ewe during the last 7 weeks of pregnancy and 4.3, 7.9 and 10.4 g per day per ewe during lactation. Lamb birth weights were 1.18, 1.77, and 1.82 kg for ewes on rations A, B, and C, respectively.

Olaloku et al. (1974) studied the influence of supplementary concentrate feeding on live weight gains and carcass quality and found that for a basal ration of grazing and hay supplemented with 1.6 and 3.2 Mcal ME (50 and 100%, respectively, of the ARC recommendations), daily gains were 57.4 and 46.5 g/animal/day, respectively. There was no difference in weight gain between sheep on 50% and those on 100% of ARC recommendations for ME. Adebambo et al. (1974) varied the dietary energy levels of six-month-old rams and found that dry matter intakes per animal were 512, 675, 764, and 875 g for the rams on the diets containing 50, 75, 100, and 125% of the ARC dietary energy recommendations for fattening (rations A, B, C, and D). There were no differences in live weight gains for the animals at slaughter time. The average live weights were 22.5, 24.2, 23.6, and 24.0 kg for animals on rations A, B, C, and D, respectively. Dressing percentages were 52.4, 50.7, 49.0, and 52.9% for animals on the respective rations.

These studies suggest that feeding above the ARC recommended energy-intake levels increased lamb birth weights and ewe milk yield. No advantages were observed from feeding growing lambs more than 50% of the ARC energy-intake recommendations.

BREEDING PERFORMANCE

The West African Dwarf ewes have an estrous cycle of about 17 to 19 days, attaining puberty by 6 months of age and sexual maturity when about 8 months old. The mean breeding weight ranges from 12.9 to 17 kg (when 6- to 13-months-old). Gestation spans 140 to 146 days for 8- to 13-month-old ewes. Birth weights for lambs of 8- to 13-month-old ewes are between 1.5 and 1.7 kg. Stillbirths are infrequent—about 3%. Dams of similar age groups have a placental weight of 0.26 to 0.48 kg. Milk production is good and weaning weight is between 6.3 and 6.5 kg.

Crosses between the West African Dwarf sheep and the Permer sheep breed, which is a cross between Blackhead Persian and Merino breeds, have breeding weights of about 15 to 21 kg and a gestation period of 142 to 147 days. Twinning frequency is about 10%. Birth weights range from 1.48 to 1.86 kg. Stillborn frequency increases to 6.7%. Placental weight is from 0.48 to 0.76 kg. Milk production is high.

Ademosun (1973) reported an average birth weight of 2.1 kg for the West African Dwarf sheep at the University of Ife farm. Lower weights were reported at the University of Ibadan, where the birth weights ranged from 1.5 to 1.7 kg. Ademosun (1973) noted that male lambs were heavier at birth and at weaning than were females. Singles were heavier at birth and weaning than were lambs of multiple births. He reported a twinning frequency of about 37% for the flock at the University of Ife.

Orji and Steinbach (1975) observed that, without a ram, it was difficult to detect estrus in the West African Dwarf sheep. They reported an estrous cycle of 17.76±0.48 days, ranging from 9 to 26 days, with estrus lasting 46 to 47 hours. They noted that 25% of the time estrus was in the morning, 15% in the afternoon, 30% in the evening, and 30% in the night. Monthly and individual differences for both estrous cycle length and duration were not significant.

The level of nutrition has been found to be related to age at puberty. Orji and Steinbach (1976a) fed roughage alone to ewe lambs, resulting in delayed puberty, reduced daily live-weight gains to puberty, and lowered body weight at puberty, compared to the performance of ewe lambs supplemented with 450 g of concentrate per day. Results of the comparisons were: mean age at puberty, 335±7.8 vs 262±16.6 days; body weight at puberty, 14.6±0.9 vs 16.0±0.7 kg; and average daily gain to puberty 29.4±4.7 vs 73.0±6.6 g (roughage alone vs supplement with concentrate). Age at first lambing ranged from 13.1 to 63.1 months, with a mean of 34.7 months, in a flock at Ado-Ekiti; whereas the range at Ibadan was from 10.9 to 37.0 months, with a mean of 22.4 months (Olaloku et al. 1974). The differences are likely due to the better feeding and management received by sheep at Ibadan.

As was observed for generation interval, the great variations that occurred from flock to flock, area to area, and time to time for the various reproductive performance parameters were primarily attributable to different feeding and management practices. Also, no selection for performance had been done and, in many cases, there was a high degree of inbreeding.

Studies of the nursing behaviour of Nigerian Dwarf ewes have noted their strong maternal relationship with the lamb, although there were cases of ewes temporarily rejecting their lambs (Orji and Steinbach 1976b). Ewes lost about 0.13 kg body weight per week, with greatest losses during the first 7 weeks of nursing. The correlation between weekly lamb weights and ewe weights was −0.87. Lamb growth was highest for the first 3 months. Lambs were weaned at 100 days of age, with an average body weight of 10.14 ± 0.38 kg. Weaning at 8 to 13 weeks has been found to reduce lamb mortality and to be otherwise suitable in the studies at the University of Ife.

DISEASES AND PARASITES

A recent study of the internal parasites of sheep on the University of Ife farm revealed that coccidia (*Eimeria* spp.) are commonly found in all age groups: adults and lambs, preweaning and postweaning. Lambs were most heavily infested (Onyema 1979). Other parasites found on the farm were, in descending order of intensity, strongyles, *Trichuris* spp., and tapeworm (*Moniezia expansa*). Strongyles were most frequent in the rainy season. A similar observation was made in Zaria in Northern Nigeria, where the highest number of oocysts were counted in July (Fabiyi 1973).

Olubajo (1974) reported that intensive farming compounded disease problems. He noted that the incidence of diseases like helminthiasis and pneumogastroenteritis increases under intensive farming conditions, unless measures are taken for strict disease and parasite control.

Akerejola et al. (1976) reported major disease problems due to gastrointestinal parasites, including subacute haemonchosis and chronic helminthiasis caused by mixed infections of *Trichostrongylus* spp., *Oesophagostomum* spp., and *Gaigeria*. Also, streptothricosis and demodicoses were mentioned as having caused severe economic losses due to their effect on skin and leather quality (Akerejola et al. 1976), especially in the north where the skins are offered for sale. In southern Nigeria, the pneumoenteritis complex was shown to be related to progressive body weight losses and eventually death.

Other major diseases are piroplasmosis, caseous lymphadenitis, foot rot, blue tongue, and nosebots (Akerejola et al. 1976). The area is also heavily infested with mosquitoes, which are reported to cause considerable distress to livestock when they attack in large numbers. Mosquitoes also act as intermediate hosts or vectors of viral, filarial, and other diseases (Hall 1977).

CONCLUSION

The West African Dwarf sheep is the dominant hair sheep breed in southwestern Nigeria and has great potential as a meat source in this region. To exploit this potential, development of sheep production will require studies of the genetic potential of the animals and of the effectiveness of cross-breeding, selection, and improved management practices at semi-intensive levels.

The level of feeding is not adequate and will require studies of the production and preservation of locally produced forages and of the agroindustrial by-products such as brewers' dried grains, cocoa husks, and rice bran. In addition to identifying the problem diseases, control measures are necessary.

To further improve sheep production in the area, better systems of marketing and distribution of both the sheep and their products will be required to produce an incentive to the farmers.

REFERENCES

Adebambo, V. O., E. A. Olaloku, and V. A. Oyenuga. 1974. Effects of variation of dietary energy levels on the growth and carcass quality of the Nigerian Dwarf sheep. *Proc. First Annual Conf. Niger. Soc. Anim. Prod.* 1(1): 104.

Adegbola, T. A., A. U. Mba, and F. O. Olubajo. 1976. Endogenous losses of nitrogen and protein requirement for maintenance of sheep. *Niger. J. Anim. Prod.* 3(1):89.

Ademosun, A. A. 1970a. Nutritive evaluation of Nigerian forages. I. Digestibility of *Pennisetum purpureum* by sheep and goats. *Niger. Agric. J.* 7(1): 19–26.

Ademosun, A. A. 1970b. Nutritive evaluation of Nigerian forages. II. The effect of stage of maturity on the nutritive value of *Stylosanthes gracilis*. *Niger. Agric. J.* 1(2): 164–173.

Ademosun, A. A. 1973. The development of the livestock industry in Nigeria—Ruminants. *Proc. Agric. Soc. Niger.* 10: 13–20.

Adeniyi, S. A. 1974. Beef cattle feeding and management in Derived Savanna. Pp. 116–123 in *Animal production in the tropics.* Ibadan: Heineman Educational Books (Nigeria) Ltd.

Adu, L. F., E. A. Olaloku, and V. A. Oyenuga. 1974. Effect of variations in energy intake during late pregnancy on lamb birth weights and lactation of the Nigerian Dwarf sheep. *Proc. First Annual Conf. Niger. Soc. Anim. Prod.* 1(1): 106–107.

Akerejola, O. O., T. W. Schillhorn van Veen, and C. O. Njoku. 1976. *Niger. J. Anim. Prod.* 3(1): 101.

Dettmers, A., C. A. Igoche, and A. Kikelomo. 1976. The West African Dwarf sheep. I: Reproductive performance. *Niger. J. Anim. Prod.* 3(1): 139–147.

Fabiyi, J. P. 1973. Seasonal fluctuations of nematode infections in goats in the savanna belt of Nigeria. *Bull. of Epiz. Dis. of Africa.* 21(3): 277–286.

Hall, H.T.B. 1977. *Diseases and parasites of livestock in the tropics.* London: Longman Group Ltd.

Loosli, J. K. and H. E. van Blake. 1974. Environment and animal productivity. Pp. 1–12 in *Animal Production in the Tropics.* Ibadan: Heineman Educational Books (Nigeria) Ltd.

Munyabuntu, C. M. 1977. Evaluation of three *Cynodon* genotypes using laboratory methods and biological studies with sheep and goats. M. Phil. Thesis, University of Ife, Nigeria.

Olaloku, E. A., V. O. Adebambo, and V. A. Oyenuga. 1974. The West African Dwarf sheep of southern Nigeria. I. Production characteristics of flocks in western Nigeria. II. The influence of supplementary concentrate feeding on weight gains and carcass quality. *Proc. Agric. Soc. Nig.* 10: 61–62

Olubajo, F. O. 1974. Pasture research at University of Ibadan. Pp. 67–78 in *Animal Production in the Tropics.* Ibadan: Heineman Educational Books (Nigeria) Ltd.

Onyema, M. A. 1979. Internal parasites of sheep on the University of Ife Teaching and Research Farm. B. Sc. Project Report. University of Ife, Nigeria.

Orji, B. I. and J. Steinbach. 1975. Studies on the biology of reproduction of the Nigerian Dwarf sheep. *Niger. J. Anim. Prod.* 2(1): 68.

Orji, B. I. and J. Steinbach. 1976a. Puberty in ewes of the Nigerian Dwarf sheep. *Niger. J. Anim. Prod.* 3(1): 97.

Orji, B. I. and J. Steinbach. 1976b. Lamb nursing in the Nigerian Dwarf sheep. *Niger. J. Anim. Prod.* 3(1): 98.

3.3

DJALLONKÉ HAIR SHEEP IN IVORY COAST

Yves M. Berger, *Centre de Recherches Zootechniques,*
République de Côte d'Ivoire

Ivory Coast has an estimated 720,000 head of sheep. Two different types of hair sheep are found all along the African Coast from Senegal to Angola: (1) a taller type usually found in the drier countries of the north, and (2) a smaller type, better adapted to the humid tropical climate, known as the Djallonké breed and also called Guinea Sheep or West African Dwarf.

This report focuses on the smaller breed, for which little information is available. Three sources are cited here: the University of Yaoundé in Cameroon (Vallerand and Branckaert 1975), the Experimental Station of Akandjé in Ivory Coast (Rombaut and Van Vlaenderen 1976), and the Centre de Recherches Zootecniques (CRZ) of Bouaké, which has been studying the Djallonké hair sheep since 1975.

After a three-year period of establishing the experimental unit and base performance levels for the breed, the Bouaké Research Center began improving the performance of the breed through management, nutrition, and selection.

DESCRIPTION OF THE BREED

Djallonké sheep are woolless; their coat consists of short, stiff hair. Their color is generally piebald black and white, with a white dominance, although some sheep are completely white or completely black. A few are piebald tan and white, and the blackbelly pattern also is found. The adult male shows a very well-developed mane of hair, 10 to 30 cm long, and has horns that make a complete spiral from rear to front. The ears are small, narrow, and horizontal. The tail is thin and small (25 cm long). The legs are generally short, giving the animal a very stout aspect. Table 3.3.1 shows some means for weight and body measurements.

TABLE 3.3.1.   WEIGHT AND BODY MEASUREMENTS OF DJALLONKÉ
                    SHEEP AT CRZ (GINISTY, 1977)

	1-year-old males	Adult females	Adult males
Weight	24 kg	23.3 kg	30 to 40 kg
Height at withers	57.7 cm	54.7 cm	50 to 60 cm
Heart girth	67.4 cm	61.9 cm	
Under-sternum height	29.5 cm	30.8 cm	
Scapulo-ischial length	60 cm		

MANAGEMENT PRACTICES IN THE AREA

Although most African villages have sheep and goats, the term management is not really applicable to their situation. Animals generally run freely
in the village, living on garbage, grasses on the roadside, and often on
crops, which is a cause for much discussion among neighbors. At night the
sheep find their way back to their owners' property. The owners have little
knowledge of breeding and nutrition and take little care of the animals.
Although each family owns a few head of sheep, their small flock is not
raised to sell for commercial purposes but is used as a bank, or source of
cash, when the owner requires extra income. Moreover, the small flock of
three to four animals demonstrates the owner's wealth and is used for gifts,
for dowries, and as sacrifices for religious purposes.

With the support of development organizations, however, some villages
are starting to follow a few management rules: they put all the sheep of a
village together with a shepherd, gather them at night in a park, and
regulate breeding seasons. These simple practices are not implemented
without difficulties. Currently, the government in Ivory Coast is trying
to increase sheep meat production and thus reduce importation. Some large
sheep farms are being developed using modern management practices. They
are generally associated with an industrial crop, such as manioc, palm trees,
etc.

MANAGEMENT PRACTICES AT THE CENTER

The Research Center is located in Bouaké, the second largest city of Ivory
Coast. The climate is tropical with well-defined rainy seasons (April-May
and August-September-October). Annual rainfall for the area is approximately 1,200 mm. The vegetation is of the arboreal-savanna type.

There are 180 ewes of the Djallonké breed at a farm 30 km north of
Bouaké housed in very primitive structures. Three groups have been

formed, each fed differently to determine the effect of nutrition on reproductive performance:

> Intensive group: 60 ewes fed on natural grass (savanna) during the day, plus a supplement based on sugarcane molasses, rice bran, and cottonseed cake throughout the year.

> Intermediate group: 60 ewes fed on natural grass (savanna) during the day, plus a supplement similar to that fed the intensive group but fed to this second group only during the productive periods (breeding, end of gestation, and during two months of lactation).

> Extensive group: control group fed only on natural grass (savanna) throughout the year, receiving no supplementation.

The supplement in grams per head per day for the intensive and intermediate groups follows:

	Rice Bran	Molasses	Cottonseed Cake	Total
Flushing (1.5 mo.)	200	50		250
Prelambing (1.0 mo.)	250	50		300
Early lactation (2.0 mo.)	250	50	50	350

The intensive group also receives the flushing supplement throughout the remaining periods of the 8-month cycle. Each group is under the care of a shepherd.

All animals are annually vaccinated against pasteurellosis and *peste des petits ruminants* (PPR). They receive a complete dipping for external parasites (ticks) once a month and are treated for internal parasites (mostly gastrointestinal strongyles) twice a year. (As trypanosomiasis is endemic to the area, but has never been diagnosed in this breed, the author assumes that the breed is tolerant to these parasites.)

From April 1976 to January 1979, ewes from the same flock had five lambings, or one lambing every 8 months. (Ewes were replaced with young ewes as needed, keeping the age composition similar in the different groups.) The lambings were in April 1976, December 1976, September 1977, May 1978, and January 1979.

REPRODUCTIVE PERFORMANCE

Age and Weight at First Lambing

In 1976, 21 ewe lambs were weaned at 140 days and put on a pasture of the legume *Stylosanthes* and supplemented daily. A ram was placed with the

TABLE 3.3.2.   DISTRIBUTION OF LAMBING OF DJALLONKÉ EWES
                    BY MONTH

Month	%	Month	%
January	8.0	July	8.4
February	7.3	August	8.2
March	7.3	September	7.7
April	8.4	October	8.6
May	7.9	November	8.8
June	9.1	December	10.3

Source: Vallerand and Branckaert (1975).

ewes, which were observed twice a day to determine the age at first estrus; the average was 259 days, with a range of 206 to 322 days. The average weight at first estrus was 16.3 kg.

Vallerand and Branckaert (1975) reported an average age at first lambing of 16.3 months (SD = 2.8) in Cameroon. Rombaut and Van Vlaenderen (1976), using a smaller sample at the Experimental Station of Akandjé, Ivory Coast, noted an age at first lambing of 11.5 months, with a range from 9.5 months to 14 months. Although the age at first lambing is very much dependent on the level of nutrition and physiological state of the ewe, the Djallonké of Ivory Coast seems to be generally early-maturing.

Sexual Cycle

The Djallonké breed can be bred throughout the year, which is a great advantage. Vallerand and Branckaert (1975) reported the distribution of 561 lambings over a period of 6 years (Table 3.3.2). Approximately the same percentages of lambings are reported for each of the different months of the year. At the Research Center in Bouaké, the average length of the estrous cycle was 17.4 days, ranging from 16 to 19 days. The length of estrus averaged 36 hours.

Length of Gestation

Mean gestation period is approximately 148.5 days. Data for the May 1978 lambings are presented in Table 3.3.3. For lambs with birth weights greater than 1.6 kg ($\bar{x}$ = 1.8 kg), the mean was 148.6 days; for those with birth weights less than 1.6 kg ($\bar{x}$ = 1.4 kg), the mean was 147.6 days.

Postpartum Anestrus

Forty-four females in lactation were exposed to a ram to determine the length of lactation anestrus. Forty-two females came back in estrus and conceived on an average of 42.4 days after lambing, with a range of 22 to 60

TABLE 3.3.3.   GESTATION PERIOD FOR DJALLONKÉ EWES AT CRZ

	Control			Intermediate			Intensive		
	No.	x̄	SD	No.	x̄	SD	No.	x̄	SD
May, 1978									
Singles	22	148.1	3.1	34	149.3	2.9	21	148.8	1.7
Twins							13	147.3	1.5

TABLE 3.3.4. LOSSES OF LAMBS IN RELATION TO INTERVAL
            BETWEEN LAMBINGS

	Interval between lambings	
Lambing	Less than 7 months	7 months and over
Lambs aborted or born dead	22%	0%
Lamb mortality the first month	30%	20%
Lamb mortality between the 1st and 6th month	41%	40%
Lamb survival after 6 months	7%	40%

Source: Rombaut and Van Vlaenderen (1976).

days. The interval between lambings for this group was 191 days. These 44 females were from the intensive group, which received a high level of nutrition. (A repetition of this trial was in progress as this report was written, and it appeared that the conception rate in the first 60 days would be lower.)

Lambing Interval

The heritability coefficient for the interval between lambings was estimated to be 0.46, using the daughter-dam regression method (Vallerand and Branckaert 1975). This high coefficient suggests that selection for short lambing interval should allow two lambings a year (although actual selection had not begun at the time of this writing). The authors found that the correlation coefficient between prolificacy and lambing interval was only −0.17. The small coefficient indicates that it should be possible to select simultaneously for the two traits to improve the number of lambs born per year and per ewe. However, Rombaut and Van Vlaenderen (1976) reported the following results within a typical village flock (Table 3.3.4).

The high mortality rates were obtained under village conditions, with little management or supplemental feeding. With better management, mortality could be decreased and the effect of lambing interval eliminated. However, these data suggest that without good flock management, a breeding program leading to intervals between lambings of more than seven months is advisable.

Prolificacy (Number of Lambs Born / Number of Ewes Lambing)

Vallerand and Branckaert (1975) reported an average of 117%, with yearly variations ranging from 107% to 120%. Ewe lambs at their first lambing had a prolificacy of 100%, ewes at second lambing, 103%, and mature ewes, 120%. Rombaut and Van Vlaenderen (1976) reported a rate of 127%. The results obtained at the Research Center of Bouaké (Table 3.3.5) are quite similar to those at Cameroon. The response in prolificacy to better nutrition has not been well determined, although there was a tendency to higher prolificacy in groups supplemented before and during the breeding season.

Lambing Rate (Number of Ewes Lambing / Number of Ewes at Breeding). With an interval of eight months between lambings, the fertility was high; values reported ranged from 94% to 96%. The results at the Research Center at Bouaké (Table 3.3.5) were generally lower because of a high abortion rate. Level of nutrition did not seem to affect fertility, but the abortion rate was much lower in the supplemented group in months when the overall abortion rate was high (19% vs 46.7% in April 1976, and 24% vs 56% in May 1978).

Fecundity (Number of Lambs Born per Year per Ewe). Averaged over three lambings (December 1976, September 1977, and May 1978), the intensive group produced 1.48 lambs per year per ewe, the intermediate group 1.39, and the control group 1.13 (Table 3.3.5). The supplemented groups and the control group differed in number of lambs born per year per ewe, but there were no significant differences between the two supplemented groups.

Because of a high mortality of 1- and 2-week-old lambs in all groups in May 1978, and in the control group in September 1977, the number of lambs weaned per ewe per year was much lower: 1.03 lambs in the intensive group, 1.02 in the intermediate group, and 0.74 in the control group. September 1977 and May 1978 lambings were in the rainy season, indicating that the rain adversely affected the nonsupplemented ewes in gestation and in lactation.

Ewe Lamb Performance

Sixty-nine ewe lambs were bred at 12 months of age in December 1977 and received the same feed as did the intensive group. Lambing occurred in May

TABLE 3.3.5. REPRODUCTION TRAITS AND LAMB MORTALITY IN THE CRZ EXPERIMENTAL FLOCK, BOUAKÉ

Year	Group	No. of ewes					No. of lambs			Conception rate (%) 3/2	Litter size 6/5	Lambs weaned per ewe exposed 8/1
		Exposed 1	Lambing 2	Conceived 3=(4+5)	Aborted 4	Lambed 5	Born 6	Born dead 7	Weaned 8			
April 76	Intens.	60	53		50%	28	29	3	26		1.04	43.3%
	Inter.	60	56			10	10	2	8		1.00	13.3
	Control	60	54			13	13	0	13		1.00	21.7
Dec 76	Intens.	70	70	66	2	64	76	2	72	94.3	1.19	102.8
	Inter.	57	56	51	1	50	58	4	50	91.1	1.16	87.7
	Control	64	64	59	2	57	63	4	58	92.2	1.10	90.6
Sept 77	Intens.	71	71	66	3	63	69	2	57	92.9	1.09	80.3
	Inter.	61	61	55	3	52	61	3	54	90.2	1.17	88.5
	Control	64	58	51	2	49	50	1	26	87.9	1.02	40.6
May 78	Intens.	69	65	60	17	43	56	0	11	92.3	1.30	15.9
	Inter.	60	58	55	12	43	48	2	11	94.8	1.12	18.3
	Control	57	57	52	30	22	23	1	5	91.2	1.04	8.8
Jan 79	All inter. Adults	159	146	137	4	133	139	2	134	95.0	1.04	84.3
Jan 79	All inter. Young & Adults	245	229	200	11	189	199	3	187	87.3	1.05	76.3

TABLE 3.3.6.   FERTILITY AND PROLIFICACY OF EWES MATED
                FIRST AT 12 MONTHS

	1st lambing	2nd lambing
Number of ewes at breeding	69	61
Number of ewes conceived (%)	62 (89.8%)	50 (82.9%)
Number of ewes aborted (%)	21 (33.9%)	5 (10%)
Number of ewes lambing (fertility)	41 (59.4%)	42 (68.8%)
Number of lambs born (litter size)	41 (1.00)	45 (1.07)

1978. They were then bred again in August 1978 for a lambing in January 1979. The results are presented in Table 3.3.6.

The abortion rate for the first lambing was very high, reducing the number of ewes lambing. A check for *Brucella melitensis* was negative. Low weight of ewes at first breeding was not the cause of abortion, since their average weight was 18 kg (SD = 2.3).

LAMB PERFORMANCE

Birth Weight

Table 3.3.7 shows the average weight of lambs observed at the Research Center in Bouaké.

Vallerand and Branckaert (1975) reported higher birth weights from ewes well fed before lambing: 2.5 and 2.3 kg for male and female singles and 2.0 and 1.8 for male and female twins. Higher birth weights usually led to a reduced mortality and a higher weight at weaning. Even more interesting were the results obtained by Rombaut and Van Vlaenderen (1976) in a study of a typical village flock in Ivory Coast. They compared the birth weight to the survival rates between 0 and 30 days and between 1 month and 5 months (Table 3.3.8). In a typical village flock with no supplemental feeding and no control of breeding season, an average of only one lamb in two survived to an age of 5 months.

Growth Rate, 0 to 30 Days

The growth rate of lambs during the first month after birth reflects the quantity of milk produced by the ewes. The average daily gain for lambs born at each lambing and for each group in the CRZ flock is shown in Table 3.3.9.

Male and female rates of growth were about the same. In December 1976, the control group did as well as did the intensive and intermediate groups,

TABLE 3.3.7.   BIRTH WEIGHT OF LAMBS BY SEASON AND LEVEL OF NUTRITION OF DAMS

Lambing	Group	Male single, kg No.	$\bar{x}$	SD	Female single, kg No.	$\bar{x}$	SD	Male & female twins, kg No.	$\bar{x}$	SD
	Intens.	31	1.8	.3	18	1.8	.2	24	1.5	.2
Dec 76	Inter.	19	1.9	.3	17	1.6	.3	16	1.5	.2
	Control	24	1.8	.3	21	1.4	.2	12	1.2	.4
	Intens.	22	1.9	.3	33	1.7	.4	10	1.6	.2
Sept 77	Inter.	16	1.9	.3	23	1.7	.3	18	1.5	.2
	Control	21	1.7	.3	26	1.6	.3			
	Intens.	17	1.8	.2	17	1.7	.4	22	1.6	.2
May 78	Inter.	17	1.6	.5	21	1.5	.3	8	1.4	.2
	Control	13	1.4	.3	8	1.7	.3			
Jan 79	All Inter.	69	1.9	.4	55	1.8	.3	12	1.4	.3

Unweighted means (singles):   All intensive   1.80
                              All intermediate 1.74
                              All control   1.60

Source: Ginisty and Berger (unpublished data).

TABLE 3.3.8.   EFFECT OF BIRTH WEIGHT ON LAMB SURVIVAL

Weight kg	% of lambs born	Mortality 0-30 days	Mortality 1-5 months
> 2	13	0	56%
1.5-2	29	0	
1-1.5	45	12%	68%
< 1	13	57%	100%

Source: Rombaut and Van Vlaenderen (1976)

probably because ewes were grazing post-fire regrowth, which has a very good nutritive value. The intensive group and the control group differed in growth rates in September 1977. In September, during the rainy season, the savanna has poor nutritive value. The intensive group and the intermediate group then showed the beneficial effect of supplementation. As a result of better nutrition during lactation and some selection, a much better average

TABLE 3.3.9. GROWTH RATE OF LAMBS FROM BIRTH TO 30 DAYS, g

Lambing date	Groups	Male single No.	$\bar{x}$	SD	Female single No.	$\bar{x}$	SD	Male & female twins No.	$\bar{x}$	SD
	Intens.	31	97	26	17	98	22	24	60	16
Dec 76	Inter.	19	92	26	16	94	15	13	62	23
	Control	23	100	21	21	91	17	11	68	17
	Intens.	23	98	30	31	92	25	9	64	23
Sept 77	Inter.	15	95	20	22	98	28	18	61	17
	Control	16	78	27	23	75	30			
	Intens.	8	97	19	3	110	16			
May 78	Inter.	6	97	25	5	61	23			
	Control (Males & females)				5	70	19			
Jan 79	All Inter.	62	131	34	51	122	24	6	83	36

Source: Ginisty and Berger, CRZ (unpublished data)

TABLE 3.3.10.   PREWEANING GROWTH RATE OF LAMBS

	Weight at 120 days	Average daily gain (ADG) 30-120 days
December 1976	12.5 kg	87 g
September 1977	11.8 kg	81 g
May 1978	9.0 kg	51 g

daily gain was obtained in January 1979. The coefficient of variability is 26% for males and 20% for females.

Growth Rate, 30 to 120 Days

In a typical village flock, Rombaut and Van Vlaenderen (1976) found an average daily gain (ADG) of 42 g between 30 and 150 days for lambs receiving no supplementation at all. Vallerand and Branckaert (1975) in Cameroon found an average daily gain of 52 g for lambs receiving no supplementation, but 64 g, 72 g, and 88 g for lambs receiving supplementation of different nutritive levels.

Berger (1979) and Ginisty (1978) reported growth rates between 30 and 120 days (Table 3.3.10); the latter age represented age at weaning. All lambs received a supplement of molasses, rice bran, and cottonseed cake. Consumption reached 50 g per day per lamb at 2 months of age and 100 g per

TABLE 3.3.11.   POST WEANING GAINS OF LAMBS

Year	No.	Length of trial days	Weight at start kg	Weight at end kg	ADG g	Feed consumption[a] g	Feed per gain kg/kg
1976	30	182	12.2	25	70	606	8.6
1977	15	90	15.4	23.8	93	690	7.4
1978	22	123	8.3	19	87	607	7.0

[a] Average feed consumption per day per animal.

day per lamb at 3.5 months. The lambs from the May 1978 lambing did not receive supplemental feeding.

Lamb Performance After Weaning

Some intensive feeding trials were undertaken to investigate the growth possibilities of the Djallonké male lambs. In 1976 and 1977, experiments started 2 months after weaning; in 1978, they started on weaning day. The feed was composed of sugarcane molasses (50% in 1976 and 1978, 40% in 1977), rice bran (25% in 1976 and 1978, 30% in 1977), and cottonseed cake (25% in 1976 and 1978, 30% in 1977). All lambs were penned. Water, mineral complex, and second-quality hay were offered ad libitum. Table 3.3.11 shows the results.

Even with a high level of nutrition, the daily growth rate of the Djallonké lamb is low (90 g); however, this trait may be improved by selection. In 1979, daily growth rate between 0 and 30 days reached 131 g for all lambs born in the intermediate group. More than 20% of the lambs weighed 14 kg or more at 90 days. Apparently, better management practices and selection can greatly improve the growth rate. The amount of feed required per kg of gain varied between 7 and 8.6 kg.

A weaning weight of 12 kg at 3 months of age and a slaughter weight of 25 kg (for male lambs) at 8 months seem to be realistic objectives.

Lamb Mortality

Mortality of lambs can be very high (Table 3.3.4) due to poor nutrition of the ewe (as in the control group for the September 1977 lambing) or due to disease (as in the May 1978 lambing, during which more than 78% of the lambs died during their first week of life). Although a disease affecting May 1978 lambs (Table 3.3.5) was not positively identified, *peste des petits ruminants* (PPR) was suspected. Several cases of this disease have been observed throughout Ivory Coast.

Average lamb mortality for the three lambings of the intermediate age

TABLE 3.3.12.   CAUSES OF LAMB MORTALITY

Causes	Mortality %	
	1966-1970	1971-1974
Born dead	3.3	4.3
Lack of milk	5	3.8
Enterotoxemia	3.3	4.5
Parasites	6.1	1
Respiratory diseases	5.8	10.1
Undetermined	7.7	.4
Preyearling mortality	32.2	24.1
Mortality of:		
Twins	44	33
Singles	29	21

Source: Vallerand and Branckaert (1975).

group has been 17.5% from birth to weaning (13.8%, December 1976; 11.5%, September 1977; and 77.1%, May 1978). In January 1979 preweaning lamb mortality dropped to 6%. Extra care for the lambs during their first weeks of life markedly reduced mortality and increased overall productivity of the Djallonké sheep. Mortality between weaning and the age of one year was nearly zero for lambs receiving adequate amounts of a well-balanced diet.

Vallerand and Branckaert (1975) in Cameroon observed a preyearling lamb mortality of 30%. Causes of mortality are given in Table 3.3.12.

The incidence of respiratory disease reflects the sensitivity of young lambs to heavy rain. A recommended practice is to avoid lambing during the rainy season or to provide some kind of a shelter. Shade and protection against rain also should be provided for weaned lambs.

Respiratory problems also are observed frequently in mature animals during the rainy season. Different kinds of pneumonia have been described, but the pneumonia due to *Pasteurella haemolytica* is the most frequently encountered. Annual vaccination against this disease is recommended in the humid tropics of West Africa.

CONCLUSION

The Djallonké sheep appear to be well adapted to a humid tropical environment, as indicated by their high fertility. This breed has a short lambing interval; 3 lambings in 2 years is common. Ewe lambs can be bred at 7 or 8 months of age. However, the Djallonké ewe has a relatively low prolificacy

(115 to 120%). Productivity is reduced by numerous abortions, relatively high adult mortality, and frequently high preweaning lamb mortality.

The growth rate of lambs after weaning is low even with reasonably good nutrition. Feed conversion efficiency also is rather low. However, some lambs show superior growth possibilities (more than 14 kg at 90 days), providing hope for an overall improvement of the breed through selection.

Research is needed to reduce lamb mortality and improve prolificacy of the breed, either by selection or by crossbreeding with a highly prolific breed adapted to a tropical environment. Also needed are low-cost nutritional supplements and improved management practices suitable to the humid tropics.

REFERENCES

Berger, Y. 1979. Rapport annuel 1978. CRZ No. 5 Zoot.

Ginisty, L. 1977. Rapport annuel 1976. CRZ No. L3 Zoot.

Ginisty, L. 1978. Rapport annuel 1977. CRZ No. 5 Zoot.

Kouadio, Kouame Bertin. 1978. L'ovin de la race Djallonké en Côte d'Ivoire: contribution à l'étude de l'amélioration des conditions d'alimentation. Mémoire de fin d'études. CRZ No. 14 Zoot.

Rombaut, D. and G. Van Vlaenderen. 1976. Le mouton Djallonké de Côte d'Ivoire en milieu villageois. Comportement et alimentation. *Rev. Elev. Med. Vét. Pays Trop.* **29** (5):157-172.

Vallerand, F. and R. Branckaert. 1975. Le race ovine Djallonké au Cameroun. Potentialités, zootechniques, conditions d'élevage, avenir. *Rev. Elev. Med. Vét. Pays Trop.* **28** (4):523-545.

SUPPLEMENT

In support and extension of the report by Y. Berger prepared with data supplied by the CRZ Station at Bouaké, one of the editors (G. E. Bradford) had the opportunity to visit the farm of SODEPALM (Société de Production de Palme) at Toumodi, between Abidjan and Bouaké. The sheep flock is under the supervision of Enrico Cesenelli, whose provision of information on the flock is gratefully acknowledged. Mean annual rainfall for the area is about 1,400 mm, with a less-defined dry season than that at Bouaké. The farm (more than 1,200 ha) is devoted principally to the production of cassava; *Stylosanthes* is grown for 2-year periods, between crops of cassava. There are also coconut trees, but coconut production is not currently emphasized. A flock of sheep is being established to utilize the *Stylosanthes* and *Brachiaria* forage, the cassava peel and waste, and the grazing available under the coconut palms.

The nucleus of the flock is a group of 250 Djallonké ewes obtained from a SODEPALM farm near Abidjan. The latter flock had been maintained continuously as a breeding unit for 12 years. The ewes had been transferred to Toumodi approximately 2 months prior to the time of Bradford's visit in April 1979.

The ewes were grazing in the coconut plantation and on *Stylosanthes* pasture for approximately 8 hours per day (600 to 1100 and 1400 to 1700) and were receiving approximately 300 g of dried cassava peel and wastes per head per day.

The sheep were in excellent condition—equal or superior to that of any group seen during this survey of hair sheep. All were entirely free of wool, with relatively short, sleek hair coats (Photo 3.1.1).

The predominant color pattern was white with black spotting; the latter ranged from a small amount of spotting on the points to predominantly black, with the second most common pattern being solid black. In addition, there were individuals with the typical blackbelly pattern, and blackbelly with varying amounts of white spotting. A small percentage of the ewes had wattles.

There was substantial variation in stature and weight among ewes of the typical Djallonké type. In addition, a few ewes indicated the possibility of some inheritance from the Sahelian long-legged breed.

A sample of 30 ewes had a mean weight of 26 kg on their arrival at Toumodi. The difference between this weight and the 23.3 kg reported by Berger is consistent with the observed differences in the condition of ewes in the two flocks.

The first 51 ewes to lamb previous to Bradford's visit had produced 31 singles and 20 sets of twins, for a mean litter size of 1.39. The lambs ranged in age from 2 to 6 weeks and only 1 of the 71 lambs had died.

Birth weights had been recorded on 19 singles and 11 sets of twins, with means of 1.96 (range 1.3 to 2.6) and 1.47 (range 1.0 to 1.9) kg, respectively.

The condition of the ewes in this flock suggests that their performance (i.e., mean mature weight of 26 kg, litter size of 1.4, and birth weight of singles of 2.0 kg) is close to the present genetic potential for the Ivory Coast strain of the Djallonké breed. Body weight might be increased further by a high-energy diet, but this would not be expected to improve productivity.

Although these values may represent current mean genetic potential, considerable variation was evident among animals in this flock, all of which were in good condition. This variation is documented by the large coefficients of variation reported by Berger and suggests the possibility of rapid changes in genetic potential via selection within the breed.

3.4

A NOTE ON CHARACTERISTICS OF HAIR SHEEP IN SENEGAL

G. E. Bradford, *University of California, Davis, U.S.A.*

Hair sheep in Senegal are of two general types: the Djallonké (forest type) found in the high-rainfall sections of the southern part of the country (latitude 13° N) and the Sahelian type in the arid northern part of the country (latitude 14° to 17° N).

The Kolda Station is located in a climatic zone similar to that described by Berger (Chapter 3.3) for the Bouaké Station in Ivory Coast, and the mean performance values for Djallonké sheep at Kolda (Table 3.4.1) are remarkably similar to those reported from Bouaké. For example, the means reported for Kolda and Bouaké, respectively, are 23.3 and 23.4 kg for mature ewe weight, 1.7 and 1.7 kg for mean birth weight, and 1.12 and 1.13 for litter size. A high preweaning mortality and a slow growth rate after 30 days have been identified as the most serious constraints to production at Kolda (Table 3.4.1). Pneumonia and dysentery were identified as the most important causes of mortality, with dysentery occurring in spite of a regular program to control internal parasites.

The hair sheep of the Sahelian zone of Senegal are considerably larger than Djallonké sheep (Table 3.4.2). When weighed for the study, the sheep were in relatively thin condition due to the sparse forage in the area. The incidence of multiple births was low, which is in agreement with data presented by Wilson (Chapter 3.5) for other Sahelian sheep.

This chapter was developed from annual reports on Senegal projects involving hair sheep. Appreciation is expressed to Dr. El Hadj Gueye, Dr. P. I. Thiongane, and Dr. I. Diallo, Centre Recherche Zootechnique, Senegal, for permission to quote from these reports.

TABLE 3.4.1.   MEAN AND STANDARD DEVIATION FOR
PERFORMANCE TRAITS OF DJALLONKÉ SHEEP,
CRZ AT KOLDA, 1978

Trait	No.	$\bar{x}$	SD
Age at first lambing, days	44	582	90
Weight at first lambing, kg	40	17.8	3.4
Lambing interval, days	336	284	72
Litter size at birth	169	1.12	.35
Perinatal mortality, %	190	4.2	−
Mortality 10 days to 6 months, %	182	45.6	−

	Males			Females		
	No.	$\bar{x}$	SD	No.	$\bar{x}$	SD
Weight, kg						
Birth	92	1.8	.5	98	1.6	.5
30 days	75	4.6	1.3	72	4.1	1.3
90 days	45	7.9	2.2	46	7.1	2.4
6 mo.	19	11.1	1.8	23	10.4	2.3
9 mo.	7	10.4	1.8	11	11.4	1.3
Mature: May (max.)	9	32.6		97	25.5	
Mature: Nov. (min.)	4	28.5		82	21.8	
Daily gain, g						
7-30 days	75	85.9	36.3	71	72.0	36.7
30-90 days	45	44.2	27.0	46	44.6	28.6

Source: Rapport Annuel d'Activité 1978
Amélioration Génétique de Mouton Djallonké au Sénégal
Centre de Recherches Zootechniques de Kolda, Sénégal.

TABLE 3.4.2.   MEAN LITTER SIZE AND BODY WEIGHT, PEUL AND
TOUABIRE BREEDS, CRZ AT DARHA, 1976-1977

	Peul			Touabire		
	No.	$\bar{x}$		No.	$\bar{x}$	
Lambs born/ewe lambing	65	1.02		28	1.11	
Lamb gain, 0-40 days, g	55	124.4		25	146.1	
Lamb gain, 40-180 days, g	39	110.6		29	92.9	
Body weight, kg[a]			SD[b]			SD[b]
Group I[c] ewes	83	32.6	3.8	–	–	–
rams	5	51.0	4.1	–	–	–
Group II[c] ewes	71	29.1	3.3	65	34.5	4.4
rams	–	–	–	5	46.4	4.4

[a] Avg. of monthly means, Jan - June 1977.

[b] Avg. of within month standard deviations.

[c] Group I animals were introduced to the flock in 1975 and were well
adapted by 1976-77; Group II animals were introduced during 1976
and still adapting to the experimental flock environment.

Source: Rapport d'Activité Année 1976-1977. Centre de Recherches
Zootechniques de Darha - Djolof, Sénégal.

SEDENTARY SHEEP IN THE SAHEL AND NIGER DELTA OF CENTRAL MALI

R. T. Wilson, *International Livestock Centre for Africa,*
Addis Ababa, Ethiopia

INTRODUCTION

This chapter draws from an interdisciplinary study of traditional livestock production systems in Central Mali in late 1977 and early 1978. The main features of the study area are shown in Figure 3.5.1. It contains about 70,000 km², with its extent being determined rather arbitrarily in relation to national boundaries and other development projects—only the southern limit is related to a natural feature. Thus, the study area cannot be considered as a natural, historical, or administrative entity; in addition, it cannot be considered the domain of a particular pastoral or agropastoral system. On the contrary, it includes a variety of natural areas which, by their characteristics and their geographical position, allow the development of a variety of agricultural cropping and pastoral systems and interacting agropastoral systems.

It is, however, possible to classify the area into three principal regions, although all three regions extend well beyond the boundaries of the study area.

Region 1 is the Live Delta of the Niger River in the east and southeast, which constitutes the present flood plain of the Niger River and its principal tributaries and effluents. Due to the annual floods, the region is devoted largely to transhumant pastoralism, although there is some rice cultivation. Cattle offer the main source of income from livestock, although some subgroups in the area own sheep, usually of the coarse-wooled Macina race.

Region 2 is the Dead Delta of the Niger River in the west central area, which is formed from an earlier flood plain of the river. The southwestern portion of this region is part of the irrigation scheme managed by the Office du Niger (since the 1930s); the principal canal of this scheme partially occupies an earlier bed of the Niger.

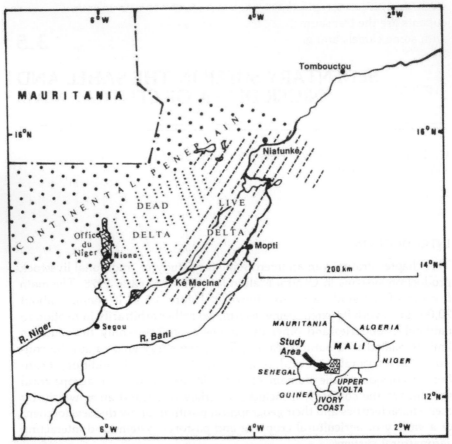

FIGURE 3.5.1 General features of the Central Mali area. A location diagram is at the lower right.

Most of the Dead Delta is sparsely populated. Livestock production is of minor importance as compared with rice production. Millet is the main subsistence crop, with some sorghum being cultivated in the south. The density of cultivation in the area decreases from south to north, although population pressure is forcing agriculture northwards into less favorable areas; a tendency to partial or total crop failures in this zone has been aggravated by droughts in the 1960s and 1970s. Oxen for draught purposes are an important part of the total livestock holdings; donkeys are extensively used for transport, and goats are generally more numerous than sheep. Some livestock are moved from the cultivated areas during the crop-growing season, but, for practical purposes, the animal husbandry system can be classed as sedentary.

Region 3 is the Continental Peneplain in the northwestern part of the study area; it has a typically sahelian climate and vegetation. Principal occupants are the transhumant/nomadic Maures, who keep cattle and sheep, with some camels and goats.

THE NATURAL ENVIRONMENT

The study area generally lies at an altitude of about 300 m, with a gentle, almost imperceptible, slope to the northeast.

Climate

The climate is generally semiarid to arid, of the dry tropical sudano-sahelian type in the south and center, changing to sahelian (saharan) in the extreme north.

Average rainfall was recorded at 570 mm at Niono, in a single season of about 50 rain-days; the standard deviation was 120 mm. Most of the rain falls in July and August—about 70% of the total—with some rainfall in June and September in most years. Rainfall at Mopti and at Ke Macina averaged 546 (131) mm and 568 (135) mm, respectively. Annual precipitation of 719 (123) mm at Segou and 323 (66) mm at Niafunke indicates the difference in the gradient from south to north. In the Live Delta, the duration and depth of the annual flood are, of course, much more important than is the amount of local rainfall.

Typical mean daily temperatures at Niono are: January, 25.8° C; May, 34.7° C; August, 27.8° C; October, 29.2° C. Diurnal variations in January may exceed 25° C, with absolute minima below 10° C (probably below 5° C in the Live Delta and in the extreme north). In the hot season of April and May, the diurnal range is generally less than 10° C, with an occasional maximum of 45° C.

Potential evapotranspiration (about 1,700 mm per year) is approximately three times annual rainfall.

Vegetation

In the Live Delta, the vegetation is typically hydrophilous grasslands except on a few scattered mounds that are not submerged in the annual inundation (and on which the permanent villages are to be found). The most productive and nutritious of these grasslands is composed mainly of *Echinochloa stagnina*, which is found in depressions known locally as *bourgou*, the name also often applied to the grass. *Vossia cuspidata* is usually found with *E. stagnina*. These best pastures are gradually being replaced by rice cultivation.

Other grasses in the flooded zone include various wild rice species

(*Oryza, Cynodon dactylon, Sporobolus spicatus*, and *Brachiaria mutica*) and sedges (*Cyperus* spp. and *Scirpus* spp.).

On the exposed mounds (*toguerre*), the palm *Borassus aethiopum* is the most abundant of the tree flora, with *Acacia nilotica* and *A. seyal* also found.

The vegetation of the Dead Delta is influenced from south to north by the variation in rainfall, the soil type, and by the time and intensity of grazing. In the sahelo-sudanian zone (to the south of the 600 mm isohyet), the field layer can be expected to produce about 2.5 t/ha of dry matter mainly from perennial grasses.

In the sudano-sahelian sector that provided the animal production information presented in this chapter, rainfall varies from 450 to 600 mm and dry matter production is about 2 t/ha from the field layer. *Pterocarpus lucens* and *Combretum micranthum* are among the principal tree species. Perennial grasses have largely disappeared in recent years due to an extended period of low rainfall and to overgrazing. The principal field layer grasses are annuals: *Schoenefeldia gracilis, Aristida* spp., *Ctenium elegans,* and *Cenchrus biflorus*. Further to the north in the sahelian sector (300 to 450 mm annual precipitation), dry matter production is probably about 1.6 t/ha, and in the saharo-sahelian sector where rainfall is below 300 mm, production is about 1.2 t/ha.

The vegetation of the Continental Peneplain is similar to that of the Dead Delta with similar rainfall, substrate, and usage.

TECHNICAL DATA ON SHEEP

Sources of Data and Evaluation Methods

During January, February, and March 1978, field studies were made in a number of villages in the vicinity of Niono. Two types of sedentary livestock systems were studied: (1) those of cultivators, primarily Bambara, whose main crop is millet; and (2) those of settlers (*colons*) of the Office du Niger who principally cultivate rice under irrigation.

Data were obtained from three sedentary Bambara villages and from three villages in the irrigated zone of the Office du Niger. In total, 43 flocks and 1,228 animals (sheep and goats) were analyzed; each of these animals was weighed and measured to provide physical data on type, sex, and age. A history of each animal, particularly that relevant to reproductivity of females, was obtained by questioning the owner, his wife, children, and herdsmen. (Age determination in small ruminants in the tropics has not been given adequate attention; experience has shown that the owners' estimates of the age of animals over 1 year old are not very accurate.)

In the field survey, sheep without permanent teeth were aged in five classes using a combination of the owners' estimates and an estimation by the author based on teeth wear. These five classes were used in establishing growth rates, but were pooled into two classes for population structure. All animals with permanent incisors were classed as having one, two, three, or four pairs of these teeth; an additional class of "aged" was used for those having very worn teeth. The latter two classes are referred to as "full mouth" throughout this text. The absolute age at which these teeth erupt is an individual characteristic subject to considerable variation; therefore animals are classified by their dentition class. Where absolute ages are required (for example, in the determination of weight for age), teeth eruptions have been arbitrarily set at 15, 21, 27, and 33 months of age, respectively, for one, two, three, and four pairs of permanent incisors.

Ownership Pattern and Flock Size

Census records in the administrative offices of Niono district relating to five Office du Niger villages indicated that 23.5% of all households owned goats and/or sheep. These initial studies were preliminary to longer-term research. To avoid antagonizing or eroding the confidence of the owners, no objective attempts were made to empirically quantify the numbers of families owning smallstock. However, it seems probable that approximately one family in every four owns smallstock, in both Bambara and Office du Niger villages.

The size of flocks founded by settlers of the Office du Niger differs substantially from those of the Bambara (Table 3.5.1). An initial sample showed that goats outnumbered sheep in the ratio of 8:3. Estimates of small stock per capita indicate 0.15 head per person for *colon* villages and 0.56 head per person for Bambara villages.

Physical Type of Sheep

Most of the sedentary sheep are of the "Sahel" type and can be ascribed to the Peul variety (Larrat et al. 1971). However, there is some evidence of out-crossing to other types, particularly to the wooled "Macina" and possibly also to the Djallonké or forest-type sheep.

In the typical Sahel type, the coat color of more than 90% of the sheep is white, occasionally with some black markings, particularly around the eyes. A few black sheep are found, along with some that are red, black pied, and red pied. The hair is generally short and fine; longer hair usually is associated with Macina breeding or (particularly in the case of colors other than white) some admixture of the long-haired Maure type. Males often carry an apron of long hair that extends from the throat down the chest and between the front legs.

About 25% of all sheep carry toggles that are variable in size and position. Horns are almost universal in males; they are slightly flattened in cross

TABLE 3.5.1.   OWNERSHIP PATTERN AND FLOCK SIZES FOR
                   SEDENTARY SMALLSTOCK

	Office du Niger villages		Bambara villages	
	Goats	Sheep	Goats	Sheep
Total number of flocks	27		16	
Flocks containing	26	15	16	9
Flocks containing goats, but no sheep	12		7	
Flocks containing sheep, but no goats		1	0	
Average flock size[a]	8.96	6.41	38.19	7.06
SD	6.03	13.51	27.75	14.81
Average flock size[b]	9.31	11.53	38.19	12.56
SD	5.87	17.00	27.75	18.27
Range in flock size	0-23	0-64	2-91	0-58

[a] For all flocks sampled.
[b] For only those flocks in which smallstock are held.

section, deeply ribbed, fit the classic "Rams horn" pattern, and are up to 65 cm in length. About 32% of the females have horns, which are up to 15 cm long, but are usually light and rudimentary. Ears are of medium length, in the range of 11 to 14 cm, and semipendulous; vestigial ears occur infrequently. The male's profile is markedly convex, while the female's is less so. The tail is thin, usually extending to just below the hocks.

The average (SD) shoulder height of 48 full-mouth females was 74.1 (4.1) cm; average live weight was 34.6 (4.9) kg; average chest girth was 80.1 (3.8) cm. No full-mouth males were found during the survey, but 2 males each with three pairs of incisors averaged 88 cm shoulder height and 51.4 kg live weight. Intensively fed castrates having three pairs of permanent incisors reached shoulder heights of 95 cm and weights of 66 kg.

Population Demography

Flock structure—estimated on the basis of a fairly small sample size—is similar to that found in traditional societies in Tchad (Dumas 1978) and in the Sudan Republic (Wilson 1976). The dominant feature in the older age groups is the preponderance of females (Table 3.5.2).

Of all ewes in the age group from 9 months old to the eruption of the first pair of incisors, 50% were pregnant, had aborted, or had given birth at least

TABLE 3.5.2. FLOCK STRUCTURES FOR SEDENTARY SHEEP

		Males		Females		Total	
Age			% total flock		% total flock		% total flock
Teeth	Months	No.	flock	No.	flock	No.	flock
Milk	Under 6	37	11.0	48	14.3	85	25.3
Milk	6-15	33	9.9	53	15.8	86	25.7
1 pair	16-21	5	1.5	43	12.8	48	14.3
2 pairs	22-27	1	0.3	32	9.6	33	9.9
3 pairs	28-33	2	0.6	33	9.9	35	10.5
Full mouth	Over 33	–	–	48	14.3	48	14.3
Total		78	23.3	257	76.7	335	100.0

Note: The analysis does not include 10 castrates in the flocks examined.

TABLE 3.5.3. NUMBER OF PARTURITIONS PER BREEDING EWE

Age (dentition)	Number of parturitions									Total
	0	1	2	3	4	5	6	7	8	
Milk teeth	9	7								16
1 pair permanent	8	29	2	3						42
2 pairs permanent		10	4	5	3					22
3 pairs permanent	1	3	6	6	4	1				21
Full mouth			9	8	13	4	2		1	35
All ewes	18	49	21	22	20	3	2		1	136

once. Therefore, half of the ewes in the 9- to 15-month age group were classified as breeding females, as were all females with one or more pairs of permanent incisors, thus accounting for 182 head or 54.3% of the flock. Eight sheep, 16.7% of the full-mouth class, were considered too old for further usefulness.

Reproductive Performance

Number of Parturitions Per Female. Information was obtained on 272 parturitions: complete life histories were known for all ewes and all data (Table 3.5.3) were included in the analysis. The mean number of parturitions per ewe was 2.0

Multiple Births and Litter Size. Data obtained on these aspects of production are summarized in Table 3.5.4. One full-mouth ewe had triplets, 0.3% of all births, and 0.7% of births in the full-mouth class. Twin births were 5.2% of all births and generally increased in frequency with increasing age

TABLE 3.5.4.   LAMBING DATA FOR 24 FLOCKS OF SEDENTARY
SHEEP

	Age of dam (expressed by means of dentition)					
	Full mouth	3 pair incisors	2 pair incisors	1 pair incisors	Milk teeth	All
Number in sample	37	21	22	42	16	138
Type of birth						
Triplet	1					1
Twin	9	2	2	1		14
Single	124	52	43	41	7	267
Total births	134	54	45	42	7	282
Total lambs born	145	56	47	43	7	298
Average litter size	1.08	1.04	1.04	1.02	1.00	1.06
Parturitions per ewe						
Mean	3.62	2.57	2.05	1.00	0.43	1.84
Mode	3	2+3	1	1	0	1
Range	2-8	0-5	1-4	0-3	0-1	0-1

(0.0, 2.4, 4.6, 3.8, and 7.3% of all births for the ewe age classes from milk teeth to full mouth, respectively). The overall figure of 5.5% multiple births is low compared with that of 14% in Sudan Republic sheep (Wilson 1976).

Death Rates. No firm information was obtained on death rates in the study area. In the Mopit area, mortality in small stock during the first year of life was estimated at 25% (Coulomb 1972) and at 10% per year thereafter. It seems probable that for lambs up to 6 months of age, mortality rates vary between 20 and 30% of all births, depending on the season and the time of year in which birth and weaning occur.

Birth Rates. Birth rates have been calculated as 4n/N where n is the number of female lambs under 6 months of age in the flock (n is then multiplied by 2 on the assumption that sex ratios are 1:1 at birth, and by 2 again to give the year's total of lambs under 6 months) and N is the calculated number of breeding females.

When n is adjusted for death rates of 20%, the derived birth rate is about 130%; for death rates of 30%, the derived birth rate is about 150%.

Lambing Interval. Based on calculated birth rate and average litter size, the lambing interval varies between 258 and 298 days, equivalent to between 1.23 and 1.42 parturitions per year.

Production Characteristics

Birth Weights. Table 3.5.5. shows the data from 111 lambs weighed within a few hours of birth (and covering the period January 1978 through June 1979).

TABLE 3.5.5.   BIRTH WEIGHTS OF LAMBS, SEX, AND TYPE
OF BIRTH

	Number	Mean birth weight kg	Standard deviation kg	Range kg
Single births				
Males	43	3.20	0.94	1.4–5.5
Females	37	3.01	0.81	1.5–5.5
Twin births				
Males	11	2.53	1.29	1.4–4.8
Females	17	2.85	0.69	1.7–3.8
Triplet births				
Males	2	2.4	–	–
Females	1	3.5	–	–
All births	111	3.00	0.90	1.4–5.5
(uncorrected for sex or type of birth)				

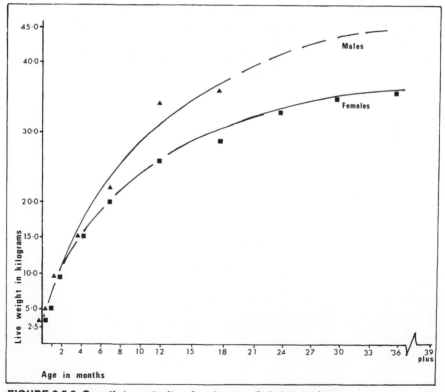

FIGURE 3.5.2 Growth to maturity of sedentary Sahel-type sheep.

Weight for Age and Growth Rates. The growth rates of lambs were determined by linear regression analysis based on estimated age and actual live weight of traditionally owned animals. The average live weight at 6 months of age (unweighted for sex) was calculated at 19.86 kg, with the relevant regression equation being $Y = 0.592X + 4.473$ (Y is weight in kilograms and X is age in weeks). Average gain in weight to 6 months of age was equivalent to 84 g per day—similar to the rate of gain of Sudan sheep in the sudano-sahelian zone of Darfur (Wilson 1976). Daily gains to 3 months of age were about 130 g.

Figure 3.5.2 shows the growth curves for males and females from birth to maturity.

REFERENCES

Coulomb, J. 1972. Projet du développement de l'élevage dans la région de Mopti (République du Mali): Etude du troupeau. Paris: IEMVT, Maisons Alfort.

Dumas, R. 1978. Etude sur l'élevage des petits ruminants du Tchad. Paris: IEMVT, Maisons Alfort.

Larrat, R., J. Pagot, and J. Vandenbushche. 1971. Manuels et précis d'élevage. 5. Manuel vétérinaire des agents techniques de l'élevage tropical. Paris: IEMVT/Ministère de la Coopération.

Wilson, R. T. 1976. Studies on the livestock of Southern Darfur, Sudan. III. Production traits in sheep. *Tropical Animal Health and Production.* 8:103–114.

4.1.1 Grade Barbados Blackbelly ram produced in Texas and used in Spurlock flock in California

RESEARCH WITH BARBADOS BLACKBELLY SHEEP IN NORTH CAROLINA

L. Goode, T. A. Yazwinski, D. J. Moncol,
A. C. Linnerud, G. W. Morgan, and D. F. Tugman,
North Carolina State University, U.S.A.

INTRODUCTION

The geography and climate of North Carolina are highly variable. The state lies at 35 to 36° N, and day length varies during the year from approximately 10.0 to 14.5 hours; there are distinct spring, summer, fall, and winter seasons. North Carolina is made up of three major geographical areas (from west to east): (1) Appalachian Mountains, (2) Piedmont, and (3) Coastal Plains.

The mountains of western North Carolina range in elevation from less than 610 m to well over 1,830 m above sea level. Summer temperatures are relatively mild and seldom exceed 29° C at the higher elevations. These mountains have ample rainfall, and cool-season forages are the main source of feed; there are no serious environmental constraints on sheep production.

In the Piedmont and Coastal Plains areas, elevations range from sea level to approximately 305 m. Summers are hot (Table 4.1.1) and relative humidity ranges from 60 to 80% much of the time. While rainfall is usually adequate, high summer temperatures reduce the performance of cool-season grasses and legumes. Warm-season forages are an important feed source. Heat stress adversely affects sheep in these areas and is a very serious problem, especially for those producers attempting to breed sheep during July and August to take advantage of high lamb prices during the spring months.

Sheep production is not a major enterprise in North Carolina; the state's approximately 10,000 brood ewes are located mainly in the Appalachian Mountain area. Hampshire, Suffolk, and Dorset are the main breeds used; crossbred Western ewes (Suffolk or Hampshire × Rambouillet) are pur-

TABLE 4.1.1.   CLIMATOLOGICAL DATA, PIEDMONT RESEARCH
                 STATION, SALISBURY, NORTH CAROLINA.
                 (4-YEAR AVERAGE)

	Month				
Item	May	June	July	August	September
Temperature, degrees C[a]					
Average high	26.4	29.9	31.3	31.3	28.3
Average low	12.4	16.5	19.1	19.1	14.9
Days over 32°C	1.5	10.5	14.0	13.3	5.0
Precipitation, cm	8.4	10.9	11.9	14.0	8.6

[a] Temperatures would be higher in the Eastern Piedmont and
Coastal Plains.

chased by some commercial producers. Most of the flocks are small and
poorly managed, and the average lamb crop marketed usually ranges from
95 to 105%. This level of production is not adequate for a viable sheep industry, since market lambs account for approximately 90% of the gross income per ewe.

AGRICULTURAL RESEARCH SERVICE FLOCK, PIEDMONT AREA

Heat Stress and Reproduction

In the Piedmont and Coastal Plains areas of North Carolina, producers seek
to have lambs born during the early winter so that they reach a market
weight of approximately 45 kg by late June. Lamb prices are usually highest
during the spring months, and hot weather and internal parasites reduce
lamb performance during the summer. Winter lambing requires that ewes
be bred during the late summer and early fall.

Heat stress is the most serious factor limiting sheep production in the
Piedmont and Coastal Plains areas. The most serious effects of heat stress
are: (1) low-quality semen and temporary sterility in rams, (2) failure of
ewes to exhibit estrus, (3) abnormal ova and fertilization failure, (4) early
embryo death, and (5) impaired fetal development.

Early embryo death and impaired fetal development are especially important when out-of-season breeding and fall lambing are required. Fall lambing of British breeds in Piedmont, North Carolina, has been characterized by
late-gestation abortions, a short gestation period (142 or 143 days), and
small, weak lambs at birth. Many lambs are born dead or die shortly after
birth. Approximately 30% of the ewes bred during the spring fail to lamb.

Research at North Carolina State University, Raleigh, has shown that when early-bred ewes were confined in a cool barn, fall lambing problems were reduced, but not eliminated. Increased production costs made confinement impractical, however. It also was apparent that hormone treatments and selection within conventional breeds offered little hope for correcting heat stress and other reproductive problems encountered in the southeastern United States. A crossbreeding program, using breeds biologically adapted to reproduce under adverse conditions, was considered to be the most promising area for further study.

Thus, a crossbreeding program was begun using two "exotic" breeds of sheep, the Finnish Landrace and the Barbados Blackbelly, to improve ewe productivity so that market lamb production could become a profitable enterprise. The Finnish Landrace originated in Finland and is a small breed noted for prolificacy and lamb vigor at birth. The Barbados Blackbelly is a small, hairy breed that apparently developed primarily from West African stock on the island of Barbados. The breed was reported to breed out of season and to be heat tolerant, hardy, and prolific. The foundation of Blackbelly sheep was obtained from several sources in Texas, Louisiana, and Mississippi; they obviously were carrying varying percentages of other breeds and may not be representative of those sheep found on Barbados.

The basic plan was to cross the Finnish Landrace and Blackbelly with other available breeds to produce a brood ewe with the following desirable traits: (1) heat tolerance, (2) out-of-season breeding and the potential to lamb regularly at 8-month intervals or less, (3) low lamb mortality at birth, and (4) adequate performance in market lambs. The next step was to evaluate the crossbred ewes for market lamb production in a terminal sire mating to Suffolk or Dorset rams.

Experiment I. This study was initiated with yearling ewes in the spring of 1971. Purebred Dorsets (D) were compared to Dorset × Blackbelly (D × B), Finnish Landrace × Dorset (L × D) and Finnish Landrace × Rambouillet (L × R) crosses. The D × B group contained ewes from reciprocal crosses of the D and B breeds (i.e., D × B and B × D ewes; the other crosses were made only one way, with the sire breed listed first). The L × R ewes were approximately 6 months old when obtained from the John Redding flock at Ashboro, North Carolina. All other ewes were produced in the university flock.

From April 15 to October 15, 1971, ewes were checked daily for estrus (using vasectomized rams) and were mated to Dorset rams for 20-day periods, alternating with 20-day periods during which they were not mated. Ewe groups grazed together on grass-legume pastures. This procedure was repeated with the same ewes during 1972, except that ewes were heat-checked only during the 20-day breeding periods and were mated to Suffolk rams. Ewes were on mixed clover-grass pasture during the grazing season.

TABLE 4.1.2.   SUMMARY OF REPRODUCTIVE PERFORMANCE OF
TWO- AND THREE-YEAR-OLD DORSET, DORSET X
BLACKBELLY, DORSET X FINNISH LANDRACE,
AND RAMBOUILLET X FINNISH LANDRACE EWES.
(EXPERIMENT I, 1971-1972)

Item	Breed group			
	D	D x B	L x D	L x R
No. ewes[a]	4	10	7	10
No. possible lambings	8	20	14	20
Avg gestation length, days	143.3^d	146.4^e	144.4^d	143.9^d
Avg lambing date	Nov. 28^d	Nov. 25^d	Jan. 15^e	Jan. 11^e
Avg litter size[b]	1.63	1.45	1.64	1.90
Ewe weight at lambing, kg	59.6^d	49.2^e	61.3^d	57.7^d
Avg lamb birth wt, kg[c]	2.4^d	3.9^e	2.9^d	3.0^d

[a] Statistical analysis based on only those ewes lambing each year.
Actual numbers per breed group were 7, 10, 9, and 10 for D,
D x B, L x D, and L x R groups, respectively.

[b] Avg. no. lambs per ewe based on total possible lambings were 1.14,
1.45, 1.39, and 1.90 for D, D x B, L x D, and L x R groups,
respectively.

[c] Least squares means, adjusted for age of ewe, sex of lamb,
type of birth.

[d,e] Means on same line with different superscripts differ
significantly (P < 0.05).

TABLE 4.1.3.   PERCENT EWES EXHIBITING ESTRUS BY PERIODS,
1971 BREEDING SEASON (EXPERIMENT I)

Item	Breed group			
	D	D x B	L x D	L x R
Ewes exhibiting estrus by May 25, %	43.0	43.0	0	0
Ewes exhibiting estrus by July 4, %	57.0	73.0	22.0	20.0
Ewes exhibiting estrus by October 15, %	100.0	100.0	100.0	100.0

TABLE 4.1.4.   SUMMARY OF REPRODUCTIVE PERFORMANCE OF
TWO-YEAR-OLD DORSET AND DORSET X
BLACKBELLY EWES (EXPERIMENT I, 1971-1972)

| Item | Breed group | |
	D	D x B
No. ewes per group	18	21
Avg gestation length, days	143.2^b	145.8^c
Avg lambing date	Nov. 29	Dec. 13
Avg litter size	1.44	1.52
Avg lamb birth wt, kg[a]	2.7^b	3.4^c
Avg ewe wt after lambing, kg	53.7^b	48.8^c

[a] Least squares means, adjusted for type of birth and sex of lamb.

[b,c] Means on same line with different superscripts differ
significantly $(P<0.01)$.

Mixed clover-grass hay or alfalfa hay was fed during the winter. Shade,
water, and salt were available continuously.

Pregnant ewes were separated from the groups 4 to 6 weeks before lambing and were fed 0.23 to 0.34 kg of ground corn per head per day. Lactating
ewes were fed 0.45 to 0.68 kg of concentrate per head per day, depending on
the quality of forage or roughage available. Results are summarized in
Table 4.1.2. The D × B ewes had significantly longer gestation periods,
heavier lamb birth weights, and weighed less after lambing than did the
other breed groups. Lambing rates based on ewes lambing each year did not
differ significantly. Dorset and D × B ewes lambed earlier $(P < 0.05)$ than
did the L × D and L × R groups.

Table 4.1.3 summarizes additional information on breeding season obtained during 1971. Ewes with Finnish Landrace breeding were found to be
much more seasonal in breeding than were the D and D × B ewes. Seasonal
breeding also has been observed in other North Carolina flocks with purebred and crossbred Finnish Landrace ewes.

In 1972, some additional D and D × B ewes were added to their respective
breed groups. The data on 2-year-old ewes lambing in 1971 and 1972 were
combined for statistical analysis; the results shown in Table 4.1.4 are
similar to those shown in Table 4.1.2. The D × B ewes had significantly
longer gestation periods, gave birth to heavier lambs, and were lighter at
lambing than were the D group. The two groups did not differ significantly
in lambing date and litter size. The correlation of gestation length with lamb
birth weight was determined for each breed group. According to the regression coefficients, increasing the gestation period by one day resulted in an

TABLE 4.1.5.   POSTWEANING GAIN OF LAMBS FROM TWO- AND
                      THREE-BREED CROSSES

Item		No. lambs	Average daily gain, kg
Two-breed crosses			
Sire breed	Ewe breed		
Blackbelly	Dorset	32	.20
Landrace	Dorset	16	.23
Landrace	Rambouillet	12	.19
Three-breed crosses			
Sire breed	Ewe breed		
Dorset	Dorset x Blackbelly	10	.29
Dorset	Dorset x Finnish Landrace	12	.29
Dorset	Rambouillet x Landrace	14	.30
Suffolk	Dorset x Blackbelly	8	.29

increase of 0.15 and 0.22 kg in the average birth weight of lambs gestated by D and D × B ewes, respectively.

Some data on the postweaning rate of gain of early weaned lambs, representing several crosses, were obtained during the spring and summer of 1972. Age of weaning ranged from 40 to 70 days. Lambs were fed clover-grass or alfalfa hay; a concentrate mixture consisting of 80% ground corn, 18% soybean meal, 1.0% ground limestone, and 1.0% salt was fed ad libitum. However, conditions were not uniform, thus the data were not analyzed statistically. The unadjusted means are shown in Table 4.1.5. The performance of F_1 Blackbelly and Finnish Landrace crosses was considered unsatisfactory for market lamb production in this area. However, when D × B, L × D, and L × R ewes were bred to a growth-line sire such as Dorset or Suffolk, lamb gain exceeded 0.27 kg per day, which was very acceptable.

The postweaning rate of gain of straight-bred Blackbelly lambs averaged approximately 0.15 kg per day. This is not adequate for market lamb production in the United States. However, these sheep could make an important contribution to the food supply in developing areas where a "walking" meat supply is needed. Blackbelly lambs appeared to have more pelvic and kidney fat than did British breeds, and the fat appeared less saturated. The fleeces of F_1 Blackbelly cross ewes were a mixture of wool and hair and ranged in weight from about 1.4 to 2.3 kg. The gestation period for a limited number of Blackbelly ewes averaged 151 days. Ewe weight after lambing averaged 41.6 kg, with a standard deviation of 6.5 kg. Twin ram and ewe lamb birth weights averaged 2.9 (SD = 0.5) and 2.8 (SD = 0.4) kg, respectively.

Experiment II. From 1974 through 1978, additional data were obtained on

TABLE 4.1.6.   PERFORMANCE OF DORSET, BARBADOS BLACK-
BELLY, DORSET X BLACKBELLY, FINNISH LANDRACE
X BLACKBELLY, AND SUFFOLK EWES BRED DURING
THE SUMMER AND FALL IN PIEDMONT, NORTH
CAROLINA (EXPERIMENT II, 1974-1978)

Item	Dorset and Blackbelly rams Ewe breed group		Suffolk rams Ewe breed group		
	D	B	D x B	S	L x B
No. ewes involved	74	48	60	82	10
No. possible lambings	141	103	142	188	18
Ewes lambing, %	82.9^a	92.2^b	91.5^x	84.6^y	94.4
Lambs born per ewe exposed	1.31	1.52	1.64	1.37	1.67
Lambs born per ewe lambing	1.58	1.65	1.79	1.63	1.77
No. lambs born	185	157	233	258	30
Lamb death loss to 30 days, %	20.0^a	6.4^b	8.2	13.2	10.0
Type of birth, %					
Singles	53.0^a	36.8^b	26.6^x	42.8^y	35.3
Twins	36.8^a	61.1^b	68.5^x	52.2^y	52.9
Triplets	9.4	2.2	5.4	5.0	11.8
Quadruplets	.8	--	--	--	--

[a,b]   D vs B ewes only; percentages on same line with different
superscripts (a,b) differ significantly (P < 0.05).

[x,y]   D x B vs S ewes only; percentages on same line with different
superscripts (x,y) differ significantly (P < 0.01).

the performance of Dorset (D), Suffolk (S), Barbados Blackbelly (B), and
Dorset × Blackbelly (D × B) ewes, as summarized in Table 4.1.6. During this
period, D and B ewes usually were assigned to two groups during the
breeding season: one group was pasture-bred to Dorset rams and the other
to Blackbelly rams, so as to obtain a combination of straightbred and
reciprocal cross lambs. The D × B and S ewes were pastured together and
were pasture-bred to Suffolk rams. Breeding began August 5 to August 10
and ended October 20 to October 25. Data obtained on a limited number of
Finnish Landrace × Blackbelly (L × B) ewes are reported in Table 4.1.6, but
were not subjected to statistical analysis. The only valid breed group comparisons in Table 4.1.6 are those between D vs B ewes and between D × B vs
S ewes.

This study showed that Blackbelly and Blackbelly-cross ewes exceeded
British breeds in percentage of ewes that lambed (P < 0.05). Lamb survival
to 30 days was significantly higher in B ewes than in D ewes and approached significance in the D × B vs S comparison. The percentage of single

TABLE 4.1.7.   MEAN RECTAL TEMPERATURES OF DORSET,
                SUFFOLK, BARBADOS BLACKBELLY, DORSET X
                BLACKBELLY, AND FINNISH LANDRACE X
                BLACKBELLY EWES IN COOL AND HOT
                ENVIRONMENTS (EXPERIMENT II)

Breed group	No. ewes	Cool env. (15.6°C, 40% R.H.)	Hot env. (35°C, 65% R.H.)
Dorset	10	39.2^a	40.4^a
Suffolk	10	39.1^a	40.0^a
Dorset x Blackbelly	10	38.8^b	39.5^b
Finnish Landrace x Blackbelly	10	38.7^b	39.3^{bc}
Barbados Blackbelly	10	38.4^c	39.0^c

a,b,c Means in same column with different superscripts differ
      significantly $(P < 0.05)$.

births was lower and twin births higher $(P < 0.05)$ in B ewes as compared to
D × B ewes. Twins are a decided asset in market lamb production, whereas
triplets and higher multiple births are usually a liability because of high
lamb death loss and poor performance.

Since the weather usually was very hot during the first 30 to 40 days of
the breeding season, data on rectal temperatures were obtained on each
breed group during cool and hot weather (Table 4.1.7). The two sets of data
were analyzed separately since different ewes were involved in these
measurements.

Significant breed-group differences were obtained when the ewes were
subjected to a cool environment. The normal rectal temperature of B ewes
was approximately 0.7° C lower than that of the British breeds. Blackbelly
cross ewes were intermediate between these extremes. This difference between B and B crosses vs British breed ewes increased under severe heat
stress. The ability of Blackbelly and Blackbelly-cross ewes to withstand the
heat stress was probably responsible for most of the superior reproductive
performance of these ewes as compared to that of British breeds in Piedmont, North Carolina.

Parasite Resistance

Some breeds of sheep appear to have natural resistance to gastrointestinal
parasites. The addition of the Barbados Blackbelly and the Finnish Landrace
to the Agricultural Research Service flock presented an opportunity to investigate parasite resistance in sheep with widely different points of origin
and genetic backgrounds. The objectives were to determine: (1) how these
breeds, or their crosses, might differ in their ability to resist gastrointestinal

parasites, and (2) what physiological mechanisms might be responsible for such resistance. The study consisted of three experiments using a total of 193 sheep.

Experiment I. Experiment I consisted of four trials and involved D, D × B, L × D, L × R, and B breed groups. All of these animals had been similarly fed and managed previous to the experiment. At the beginning of each trial, the sheep were treated with standard doses of an anthelmintic. Then the breed groups, within trials, were grazed together on grass-legume pastures that previously had been contaminated by grazing of sheep carrying a mixed infection of parasites. Grazing periods ranged from 70 to 98 days and anthelmintics were not administered during grazing. Parasite ova were not found in the feces for at least 21 days after treatment, attesting to the efficacy of the anthelmintics.

Fecal samples were examined for parasite ova according to a modified McMasters technique. Breed groups were compared on the basis of fecal egg counts expressed as eggs per gram of feces (EPG). Although fecal egg counts were made at frequent intervals, only the final EPGs were analyzed statistically. At the final fecal examination of trials 1, 2, and 3, ova were counted and stratified into the following groups: (1) *Haemonchus contortus* and *Oesophagostomum* spp. (HO group); (2) *Cooperia* spp., *Tricho-strongylus* spp., and *Ostertagia* spp. (CTO group); and (3) *Nematodirus* spp.(NEM). The results are summarized in Table 4.1.8.

These data clearly show that D × B sheep had lower CTO and total EPG levels than did the other breed groups (P < 0.05). In trial 1, differences between the D, L × D, and L × R groups were not significant for these parameters.

The relationship of host hemoglobin type and EPG levels also was determined in trial 1. Several authors have reported that sheep with type A and AB hemoglobin are more resistant to gastrointestinal parasites than are those with type B hemoglobin. In the study reported here, however, D and D × B ewes, which had the highest and lowest total EPG levels, were all of type B hemoglobin. Thus, hemoglobin type was not related to the differences in EPG levels observed in these groups. The L × D and L × R ewes were divided evenly between B and AB type hemoglobin. In these groups, those with AB hemoglobin had lower EPG levels than did those with type B hemoglobin. However, the groups had relatively few sheep and the differences in levels were not significant.

Fecal egg counts for HO parasites did not vary significantly in any of the trials, and NEM levels were not consistent among trials.

Experiment II. Breed differences shown in Experiment I could have been due to a lethal or inhibitory effect that resulted in fewer fecund parasites in D × B and B animals, or to differences in grazing habits so that these sheep ingested fewer infective larvae. Experiment II was designed to investigate

TABLE 4.1.8.   MEAN FINAL EGGS PER GRAM OF FECES LEVELS
(EXPERIMENT I AND II)

Item	Breed groups				
	D x B	D	L x D	L x R	B
Experiment I					
Trial 1 (yearling ewes)					
No. of animals	11	7	9	10	--
Parasite group					
CTO	236^a	1871^b	2078^b	1470^b	--
HO	0	14	0	0	--
NEM	0^a	200^b	0^a	20^a	--
Total	236^a	2085^b	2078^b	1490^b	--
Trial 2 (ram lambs)					
No. of animals	13	11	--	--	--
Parasite group					
CTO	3038^a	5473^b	--	--	--
HO	154	209	--	--	--
NEM	38	64	--	--	--
Total	3231^a	5745^b	--	--	--
Trial 3 (ewe lambs)					
No. of animals	19	26	--	--	--
Parasite group					
CTO	563^a	1162^b	--	--	--
HO	195	804	--	--	--
NEM	0	23	--	--	--
Total	758^a	1988^b	--	--	--
Trial 4 (mature ewes)					
No. of animals	18	18	--	--	18
Total EPG	83^a	707^b	--	--	29^a
Experiment II					
Trial 1 (ram lambs)					
No. of animals	4	4	--	--	--
Parasite group					
CTO	1675^a	4450^b	--	--	--
HO	1550	1300	--	--	--
NEM	0	0	--	--	--
Total	3225^a	5750^b	--	--	--
Trial 2 (ewe lambs)					
No. of animals	4	4	--	--	--
Parasite group					
CTO	325^a	4900^b	--	--	--
HO	1125	9625	--	--	--
NEM	0	0	--	--	--
Total	1450^a	14525^b	--	--	--

a,b Means on the same line with different superscripts differ
significantly ($P < 0.05$).

TABLE 4.1.9.   MEAN NUMBERS OF PARASITES RECOVERED
AND EGGS PRODUCED PER PARASITE
(EXPERIMENT II)

Item	Breed group	
	D x B	D
Parasites recovered		
Trial 1 (ram lambs)		
No. of animals	4	4
Parasite group		
CTO	18925	20590
HO	535	530
NEM	5^a	80^b
Total	19465	21200
Trial 2 (ewe lambs)		
No. of animals	4	4
Parasite group		
CTO	7615	10850
HO	950	3115
NEM	0	20
Total	8565	13985
Eggs produced per parasite		
Trial 1 (ram lambs)		
No. of animals	4	4
Parasite group		
CTO	$.087^a$	$.224^b$
HO	2.034	2.956
Trial 2 (ewe lambs)		
No. of animals	4	4
Parasite group		
CTO	$.059^a$	$.439^b$
HO	$.647^a$	2.905^b

[a,b] Means on the same line with different superscripts differ
significantly (P < 0.05).

these possibilities. The experiment consisted of two trials in which 8-month-old lambs were treated with anthelmintics so that EPG levels were reduced to zero. Lambs within trials were then reinfected with standard numbers of larvae and were fed and maintained on concrete surfaces throughout the experiment.

In trial 1, four D and four D × B ram lambs were infected with approximately 60,000 larvae. Four D and four D × B ewe lambs were used in trial 2 and each animal was infected with approximately 35,000 larvae. The rams and ewes were slaughtered at 48 and 49 days after infection. Parasites were collected from the abomasum, small intestine, and large intestine and were counted and stratified as to HO group, CTO group, and NEM.

Table 4.1.9 summarizes the results of the final fecal egg counts. These findings support those of Experiment I: parasites of the CTO group were mainly responsible for the significantly lower EPG levels in D × B sheep. These results also show that differences in EPG levels among breed groups were not due to differences in grazing habits and ingestion of fewer larvae. These animals had been reinfected with the same number of larvae, were maintained on concrete, and were not allowed to graze.

Data on numbers of parasites recovered from the digestive tracts are summarized in Table 4.1.9. While NEM parasites were lower (P < 0.05) in D × B rams, no significant differences were found in number of CTO and HO parasites on either rams or ewes. On the other hand, when egg production per parasite recovered was determined (Table 4.1.9), CTO parasites were shown to produce fewer ova (P < 0.05) in D × B hosts in both trials. Egg production per HO parasite was significantly lower in D × B ewes, but not in rams. Thus, parasites in D × B hosts were less fecund, and this, rather than a reduction in number of parasites, appears to be the primary cause for breed variation in EPG levels. It is also possible that the survival and/or function of male and female parasites varied with breed group.

Experiment III. This study was designed to compare physiological responses in animals carrying a mixed infection of parasites with those of animals carrying a relatively pure *H. contortus* infection. The experimental animals were mature, nongravid D, B, and D × B ewes that had acquired a mixed infection while grazing contaminated grass-legume and small-grain pastures and were then maintained on concrete throughout the study.

The experiment consisted of three periods. In period 1, data were obtained while the ewes were carrying the worm burdens acquired during grazing. Period 2 was a transition period during which information was obtained after the ewes were treated with an anthelmintic. In period 3, each animal was reinfected with approximately 10,000 larvae estimated to be 98% *H. contortus*. Data were obtained on the following: EPG levels, hemoglobin levels (Hb, g/100 ml), mean corpuscular hemoglobin concentrations (MCHC) and white blood cell levels (WBC, periods 2 and 3 only). Specific antibody titers were also measured in serum. The technique was one of indirect hemagglutination wherein tanned red blood cells from the sheep were sensitized with an antigen prepared by homogenizing third-stage *H. contortus* larvae in buffered saline. The microtitration technique was used to determine the actual titer. The results are summarized in Table 4.1.10.

In period 1, B and D × B ewes had significantly lower EPG levels than did D ewes when removed from pasture on March 6 and after 8 days on concrete (March 14). These data are in agreement with those obtained in Experiments I and II with sheep carrying mixed infections of parasites. There was a marked rise in all EPG levels from March 6 to March 14. A logical explanation is that removing ewes from exposure to infective larvae on March

TABLE 4.1.10. MEAN EGGS PER GRAM OF FECES, HEMOGLOBIN, MEAN CORPUSCULAR HEMOGLOBIN CONCENTRATION, WHITE BLOOD CELL, AND ANTIBODY LEVELS (EXPERIMENT III)

| Item | | Breed groups | | |
		B	D x B	D
No. of animals		5	6	6
Period 1				
EPG	March 6	440^a	917^a	4000^b
	March 14	2430^a	2300^a	5667^b
	March 24	1830	3083	4433
Hb	March 6	11.14^a	10.30^{ab}	8.55^b
	March 21	10.64^a	9.55^a	7.63^b
MCHC	March 6	.346	.336	.334
	March 21	$.353^a$	$.339^{ab}$	$.333^b$
Antibodies	March 14	14.40	38.67	40.67
Period 2^c				
EPG	March 28	0	0	0
	April 18	500	70	0
Hb	April 2	10.92	10.55	9.58
	April 25	11.44	10.73	9.60
MCHC	April 2	.352	.341	.336
	April 25	.358	.335	.342
WBC	April 2	10720	8858	7675
	April 25	10960	10093	8825
Period 3^d				
EPG				
Preinfection	May 2	140	30	150
Final	July 1	860	1000	2220
Hb				
Preinfection	May 2	11.08	10.63	9.28
Final	June 24	10.52	10.85	9.27
MCHC				
Preinfection	May 2	.352	.340	.345
Final	June 10	.349	.342	.335
WBC				
Preinfection	May 2	10110^a	9550^a	6800^b
Final	June 24	10460	10133	7133
Antibodies				
Preinfection	May 6	7.2	9.3	13.3
Final	May 24	22.4	18.7	21.3

[a,b] Means on same line with different superscripts differ significantly (P < 0.05).
[c] Animals drenched March 26.
[d] Animals drenched April 26 and reinfected May 6.

6 reduced the immune inhibition of fourth-stage larvae so that maturation of inhibited larvae present in the digestive tract occurred in the sheep used in this study.

On March 6, B ewes had a higher Hb level (P < 0.05) than did D ewes. The D × B group was intermediate between the parent breeds. By March 21, both B and D × B had significantly higher Hb levels; and Hb loss, which measures infection severity, was 0.5, 0.75, and 0.92 g/100 ml for B, D × B, and D ewes, respectively. Blackbelly ewes also had a higher (P < 0.05) MCHC level than did D ewes on March 21. The B and D × B ewes were more resistant to a mixed infection of parasites than were D ewes based on EPG, Hb, and MCHC levels in period 1.

Differences in EPG, Hb, MCHC, and WBC levels did not vary significantly among breed groups during period 2. Only initial and final values are shown in Table 4.1.10. At the beginning of period 3, EPG, Hb, MCHC, WBC and antibody levels were obtained before the animals were reinfected with *H. contortus* larvae. These data and final values for these variables are presented in Table 4.1.10.

Differences in EPG levels did not vary significantly among breed groups at any time after infection. These results are in marked contrast to those obtained when these same ewes were carrying a mixed infection during period 1. They do agree with previous data from Experiments I and II in that EPG levels for HO parasites did not vary significantly among breed groups.

Preinfection WBC levels were higher (P < 0.05) in B and D × B animals, but not during other sampling methods. Breed differences were not significant for Hb and MCHC levels, and these data do not agree with those of period 1. Reasons for breed differences obtained when sheep were carrying mixed infections vs a relatively pure *H. contortus* infection are not apparent.

Antibody titers obtained during Experiment III were highly variable and breed differences were not significant at any time. There was no evidence that circulatory antibodies were involved with parasite resistance in this study.

APPALACHIAN MOUNTAIN RESEARCH STATION FLOCK

Dorset × Blackbelly, Dorset × Finnish Landrace, Rambouillet × Finnish Landrace, and grade Suffolk ewes were compared in an accelerated lambing study conducted at the Upper Mountain Research Station located in the mountains of western North Carolina. The elevation at this station is approximately 914 m above sea level, and summers are relatively cool.

All ewe breed groups were pasture-bred to the same Suffolk rams in September and October 1973. Thereafter ewes were exposed to rams beginning approximately 30 days after each lambing in an attempt to produce

TABLE 4.1.11. PERFORMANCE OF DORSET X BLACKBELLY, DORSET X FINNISH LANDRACE, RAMBOUILLET X FINNISH LANDRACE, AND GRADE SUFFOLK EWES IN AN ACCELERATED LAMBING STUDY

Item	Breed group			
	D x B	L x D	L x R	Suffolk
First lamb crop (Feb. 5, 1974)[a]				
No. ewes/no. ewes lambing	15/15	15/15	15/15	16/16
No. lambs born per ewe[b]	1.87	2.13	2.47	1.88
No. lambs marketed per ewe[b]	1.73	1.73	2.00	1.69
Lamb market wt, kg	44.4	44.9	45.4	46.7
Age at marketing, days	158	160	163	141
Lamb wt per day of age[c], kg	.28	.29	.28	.34
Second lamb crop (Sept. 13, 1974)				
No. ewes/no. ewes lambing	16/15	16/14	16/10	15/9
No. lambs born per ewe	1.75	1.63	1.13	.93
No. lambs marketed per ewe	1.44	1.44	1.00	.73
Lamb market wt, kg	44.4	45.4	44.4	43.5
Age at marketing, days	169	174	177	175
Lamb wt per day of age, kg	.26	.26	.25	.25
Third lamb crop (Mar. 13, 1975)				
No. ewes/no. ewes lambing	16/15	16/15	16/12	15/13
No. lambs born per ewe	1.94	2.25	1.69	1.67
No. lambs marketed per ewe	1.69	1.56	1.50	1.40
Lamb market wt, kg	44.0	44.0	44.4	44.9
Age at marketing, days	165	161	153	131
Lamb wt per day of age, kg	.27	.27	.29	.34
Fourth lamb crop (Oct. 15, 1975)				
No. ewes/no. ewes lambing	16/10	16/0	15/0	16/0
No. lambs born per ewe	.94	–	–	–
No. lambs marketed per ewe	.88	–	–	–
Lamb market wt, kg	44.4	–	–	–
Age at marketing, days	164	–	–	–
Lamb wt per day of age, kg	.27	–	–	–

[a] Average lambing date.
[b] No. lambs based on total number ewes per breed group.
[c] Includes lamb birth weight.

lambs at 8-month intervals or less. Orphan lambs and one member of each set of triplets were allowed to nurse colostrum and were then placed in a nursery and reared artificially. In the first lamb crop, lambs weighing 13.6 to 15.9 kg were weaned from one-half of the ewes in each breed group and were finished in drylot. The remaining lambs, and lambs in subsequent lamb crops, were left on their dams and were creep fed until they reached a market weight of about 45.36 kg. Thus, ewes were rebred in lactation except for those in the early weaned group of the first lamb crop.

The results of this study are partially summarized in Tables 4.1.11 and

TABLE 4.1.12.   SUMMARY OF PERFORMANCE OF DORSET X
BLACKBELLY, DORSET X FINNISH LANDRACE,
RAMBOUILLET X FINNISH LANDRACE, AND GRADE
SUFFOLK EWES BRED FOR FOUR LAMB CROPS IN
AN ACCELERATED LAMBING STUDY[a]

| Item | Breed groups | | | |
	D x B	L x D	L x R	Suffolk
Avg no. ewes per breed group	15.75_c	15.75_{cd}	15.50_d	15.50_d
Avg no. ewes lambing	13.75^c	11.00^{cd}	9.25^d	9.50^d
No. of lambs born per ewe	6.48	5.97	5.29	4.45
No. of lambs marketed per ewe	5.71	4.70	4.52	3.81
Lamb weight per day of age, kg[b]	$.27^c$	$.27^c$	$.27^c$	$.31^d$
Lamb age at market wt, days	164.0^c	165.0^c	164.0^c	149.0^d
Kg lamb marketed per ewe	253.8_c	211.1_d	202.8_d	171.1
Ewe weight at lambing, kg	56.7^c	72.1^d	67.1^d	85.3^e
Wt lamb marketed per unit weight ewe	4.48^c	2.93^d	3.02^d	2.00^d

[a]   Data are summarized over a 26-month period (September 1, 1973 to
November 1, 1975).

[b]   Includes lamb birth weight.

[c,d,e]   Means on same line with different superscripts differ significantly
($P < 0.05$ except for avg no. ewes lambing, where $P < 0.10$).

4.1.12. A surprising percentage of ewes rebred during lactation in the first
three lamb crops. This trait is highly desirable and would greatly reduce the
amount of concentrate feed required for accelerated lamb production.
However, no D × L, R × L, or S ewes rebred after lambing in March 1975,
whereas 63% of the D × B ewes lambed in the fall of 1975. When the four
lamb crops are combined, the proportion of ewes lambing in the D × B
group was higher ($P < 0.10$) than in the other breed groups. These tables
show the superiority of the D × B ewes in ability to breed and lamb out of
season. D × B ewes lambed at a 6.5-month interval over a 26-month period.
This shortening of interval between lamb crops increased lamb production
markedly.

Suffolk lambs gained faster and reached market weight earlier ($P < 0.05$)
than did lambs from other breed groups; there were no significant differences in these traits among other breed groups.

The most important trait in Table 4.1.12 is weight of lamb marketed per
unit of ewe body weight. The D × B ewes excelled in this respect. Previous

TABLE 4.1.13.   LEAST-SQUARES MEAN FOR AVERAGE DAILY GAIN, FIRST LAMB CROP OF ACCELERATED LAMBING STUDY

| | Average daily gain, kg | | |
Item	Weaned lambs	Nonweaned lambs	Breed group average
Dorset x Blackbelly ewes	.25	.26	.26[a]
Finnish Landrace x Dorset ewes	.28	.24	.26[a]
Rambouillet x Landrace ewes	.25	.27	.26[a]
Grade Suffolk ewes	.31	.28	.29[b]
Overall	.27	.26	
Artificially reared lambs	.28	--	--

[a,b] Means in same column with different superscripts differ significantly (P < 0.05).

TABLE 4.1.14.   SUMMARY OF LAMB FEED EFFICIENCY, FIRST LAMB CROP OF ACCELERATED LAMBING STUDY

Item	Kg of feed per kg of gain[a]
Early weaned lambs	
Concentrate	3.03
Hay	1.73
Total	4.76
Nonweaned lambs	
Concentrate	1.23
Hay	.09
Total	1.32
Artificially reared lambs[b]	
Concentrate	2.91
Hay	1.99
Total	4.90

[a] Does not include feed consumed by ewes.
[b] Does not include milk replacer.

studies have shown that, while the D × B ewe weighs considerably less (P < 0.05) than the other ewes, she lambs without serious problems when bred to Suffolk rams, she is a good milker, and her lambs are vigorous at birth.

Table 4.1.13 summarizes the data on lamb daily gain for the first lamb crop. As expected, Suffolk lambs gained more rapidly than did the others (P < 0.05), but differences between the D × B, L × D, and L × R groups and between weaned and nonweaned lambs were not significant. Table 4.1.14

summarizes feed per lamb and shows that the practice of leaving lambs on the ewe has a decided advantage over early weaning.

SUMMARY

In Piedmont, North Carolina, the reproductive performance of Barbados Blackbelly (B) and/or Dorset × Blackbelly (D × B) ewes was superior to that of the Dorset (D), Suffolk (S), Finnish Landrace × Dorset (L × D), and Finnish Landrace × Rambouillet (L × R) ewes. The D × B ewes had longer gestation periods, heavier lambs at birth, and weighed less after lambing than did the D, D × L, and R × L groups. Differences in litter size were not significant. The D and D × B ewes were earlier in estrus and lambing (P < 0.05) than were the L × D and L × R ewes.

In a second experiment, B ewes outperformed D ewes in percentage of ewes lambing and in lamb survival to 30 days of age (P < 0.05). The percent of ewes lambing was higher (P < 0.05) for D × B ewes than for S ewes, and lamb survival to 30 days was also substantially higher. The percentage of single births was lower and that of twin births higher (P < 0.05) for B and D × B ewes, than for D and S ewes. Rectal temperatures of D and S animals were significantly higher than those of B and D × B animals under both normal and heat-stress conditions.

Comparisons of fecal egg counts, of hemoglobin levels, and of mean corpuscular hemoglobin concentrations showed that B and D × B sheep were more resistant to gastrointestinal parasites than were D, L × D, and L × R sheep. Differences in fecal egg counts were due to lower (P < 0.05) egg production by *Cooperia* spp., *Trichostrongylus* spp., and *Ostertagia* spp. in D × B hosts. Breed differences were not related to ewe hemoglobin type, and there was no evidence from this study that circulating antibodies were involved with parasite resistance.

D × B, L × D, L × R, and grade Suffolk ewes were compared in an accelerated lambing study conducted over a 26-month period in the Appalachian Mountain area of North Carolina. When the four lamb crops of the study were combined, the proportion of ewes lambing in the D × B group was higher (P < 0.10) than that in the other groups. The D × B ewes lambed at a 6.5-month interval. Differences in litter size were not significant. Suffolk lambs gained faster and reached market weight earlier (P < 0.05) than did other lambs. The D × B group had the highest (P < 0.05) weight of lamb marketed per unit of ewe body weight.

4.2

THE ST. CROIX SHEEP
IN THE UNITED STATES

Warren C. Foote, *International Sheep and Goat Institute,*
Utah State University, U.S.A.

The first St. Croix sheep introduced into the United States were imported from the Island of St. Croix in the 1960s by Michael Piel of Maine. He imported fewer than 10 animals for use in crossbreeding; no descendants of pure St. Croix breeding remain.

In 1975, another group of St. Croix sheep was imported by the International Sheep and Goat Institute, Utah State University, Logan. The author selected 22 ewes and 3 rams to be taken to Utah in June 1975. No production records were available on the animals prior to importation. The criteria used in making selections were white color, lack of wool (as far as possible), and average or better body size and general conformation. Younger animals were selected to provide for a longer production period after importation. Animals were selected from 3 flocks that were chosen because the sheep appeared to be the most homogeneous, or pure, as indicated by color, size, conformation, and freedom from wool.

Many of the ewes were pregnant when purchased; thus records of some production parameters were begun on their arrival in Utah. The major initial effort was to increase flock numbers as rapidly as possible. Ewes were exposed for breeding throughout the year.

St. Croix sheep are very tractable and easy to handle; they are active and vigorous, but show no tendency to be wild. The males are very active breeders. They have adapted to the severe and variable climatic conditions of Utah very well, growing a very heavy winter coat of mixed wool and hair that is shed in the spring.

These St. Croix sheep were brought to the United States to: (1) measure their reproduction and production performance, and (2) determine their usefulness as a pure breed, or through crossing, in increasing sheep production in appropriate geographic and climatic areas of the United States (and in other countries of the world including the Middle East and Africa). This flock is considered a separate breed of hair sheep and is being established as the St. Croix breed. Reproduction and production standards are being developed and will be used by the International Sheep and Goat Institute to characterize the breed. None of the sheep will be released for private or commercial use until this task is completed and flock numbers have become sufficient to warrant such an expansion.

In 1976, as a part of the North Central Regional Research Project (NC-111), a cooperative research program was arranged for St. Croix sheep to be taken to the Ohio Agricultural Research and Development Center, Wooster, Ohio, under the direction of Charles F. Parker, and to the University of Florida, Gainesville, under the direction of Phillip E. Loggins. In December of 1976, approximately 30 ewes were divided into 3 groups by age and reproductive performance: one group was sent to Ohio, another to Florida, and another was kept in Utah. In addition, 5 to 6 rams were sent to Ohio and to Florida. The research objectives included measurement of the production and reproduction of purebreds and crosses in different geographic locations. In this cooperative work, 6-month lambing intervals were established in Ohio and Utah by allowing a 40-day breeding period beginning August 1 and February 1. Each year, the ewes in Florida were exposed for breeding for a 40-day period beginning on July 15. Lambs at all 3 locations were weaned when 60 days old. Sire lines were established to prevent inbreeding.

In 1978 a small group of St. Croix (5 rams and 3 ewes) was taken to California State Polytechnic University, Pomona, for research under the direction of Edward A. Nelson, with primary focus on male reproductive traits. During early 1979, 3 rams were taken to Texas A&M University, San Angelo, for crossbreeding research with other genotypes of hair sheep. This research was directed by Maurice Shelton.

The following preliminary data provide estimates of reproduction and production performance of the St. Croix and of some crosses between the St. Croix and other breeds.

In Utah, the mean body weight of mature ewes and rams (3 years and older) was 54 and 74 kg, respectively (Table 4.2.1). In Florida, the mean weight for *all* ewes in the breeding flock was 36 kg (Table 4.2.2).

Tables 4.2.3 to 4.2.6 summarize the reproductive parameters at the three locations. Fertility and prolificacy varied among locations and probably were influenced by lambing frequency. In Utah, prolificacy was 2.00 to 2.50 for ewes bred at one year and older; in Ohio, prolificacy ranged from 1.50 to 2.12 for ewes bred at that age. Prolificacy was lowest in Florida (1.40 to 1.50), but this measurement included ewes of all ages. In Utah, prolificacy of lambs and yearlings was 1.60 and 1.75, respectively.

In Florida, fertility was 77 to 90%, within an annual lambing schedule.

TABLE 4.2.1.   MEAN BODY WEIGHTS OF ST. CROIX, ST. CROIX X RAMBOUILLET, AND RAMBOUILLET LAMBS AT BIRTH AND OF MATURE ST. CROIX EWES AND RAMS, kg

| | Sex | | | | | | Type of birth | | | | | | | | | | | | | | | |
| | Male | | | Female | | | Single | | | Twin | | | Triplet | | | Quadruplet | | | Total | | |
Genotype	No.	x̄	+SE	No.	x̄	+SE	No.	x̄	+SE	No.	x̄	+SE	No.	x̄	+SE	No.	x̄	+SE	No.	x̄	+SE
St. Croix x St. Croix																					
Birth	72	2.86	+.06	71	2.63	+.08	27	3.13	+.12	81	2.68	+.07	30	2.70	+.13	8	2.12	+.10	146[a]	2.74	+.17
Mature 3 yrs & older	2	74.0	+1.86	12	54.4	+3.58	--			--			--			--			--		
St. Croix x Rambouillet																					
Birth	12	4.54	+.12	10	3.71	+.24	8	4.45	+.12	14	3.98	+.22	--			--			22	4.16	+.28
Rambouillet x Rambouillet																					
Birth	9	5.25	+.30	10	4.65	+.44	11	5.23	+.41	8	4.69	+.34	--			--			19	4.94	+.49

[a] Sex of 3 animals was not recorded.

Source: R. C. Evans, A. J. Svejda, and W. C. Foote, Utah State University, unpublished data, 1979.

TABLE 4.2.2.   MEAN EWE BODY WEIGHTS AND HEMOGLOBIN (Hb)
                  LEVELS FOR 1977-79

Items	Breed or cross (1977-78)			Breed or cross (1978-79)		
		St. Croix x			St. Croix x	
	St. Croix	Florida Native	Florida Native	St. Croix	Florida Native	Florida Native
Number of ewes	9	45	45	13	49	50
Number of ewes lost	0	2	0	1	1	2
Hb level[a]	9.5	8.4	8.2	9.8	9.9	9.8
Weight, kg[b]	38.7	41.4	42.2	36.2	39.2	39.4

[a] Mean of Hb levels taken on 8/5/77, 11/5/77, and 2/23/78 for 1977-78
     and 8/10/78, 11/21/78, and 2/23/79 for 1978-79.

[b] Mean of ewe body weights taken on 8/5/77, 11/17/77, and 2/23/77
     for 1977-78 and 8/9/78, 11/20/78, and 2/26/79 for 1978-79.

Source: P. E. Loggins, University of Florida, unpublished data, 1979.

When exposed twice a year at Utah (excluding ewe lambs) and Ohio (excluding ewe lambs and yearlings), the percentage of ewes lambing in any one mating period varied from 25 to 100%.

These values must be regarded as preliminary because the numbers of seasons and numbers of lambings were limited; the twice-a-year lambing group, for example, involved only 3 intervals. Ewes may require more time to adjust to an accelerated program, and appropriate management programs must be developed.

Table 4.2.4 shows the response of ewes to twice-a-year lambing in Utah, measured as the proportion of ewes lambing consecutively for the period observed. The percentage of ewes that lambed at 6-month intervals for 3 consecutive intervals was 50% for ewes that were 1 year old or older when first exposed. None of the ewes that began exposure at less than 1 year old lambed for 3 consecutive 6-month lambing intervals. In Utah, the mean overall lambing interval for all ewes measured on a 6-month interval was 233 days, with the range from 185 to 359 days. It seems probable that the proportion of ewes lambing consecutively can be influenced by shifting the breeding periods to further minimize interference from the effects of seasonal anestrus.

The mean gestation length for 34 ewes was 149.9 days (Table 4.2.3). The age at which ewes demonstrated first or puberal estrus was not measured; however, each ewe was checked for estrus during the breeding period in

TABLE 4.2.3.   REPRODUCTIVE PARAMETERS FOR ST. CROIX AND RAMBOUILLET EWES PLACED ON 6-MONTH LAMBING INTERVALS IN UTAH[a]

Breed of ram	Breed of ewe	No. ewes exposed	Ewes in estrus No.	%	Ewes lambing No.	%	No. lambs born	Lambing rate Lambs born/ ewe exposed	Lambs born/ ewe lambing	Live lambs at birth No.	%	Weaning rate[b] No.	%	Per ewe lambing	Per ewe exposed
August 1977 - January 1978[c]															
St. Croix x St. Croix															
	6 months[d]	8	6	75	5	62	8	1.00	1.60	8	100	4	50	.50	.80
	1 yr and older	13	12	92	11	85	27	2.08	2.45	24	89	17	63	1.31	1.54
February 1978 - July 1978															
St. Croix x St. Croix															
	6 months	0	-	-	-	-	-	-	-	-	-	-	-	-	-
	1 yr and older	21	18	86	13	62	26	1.14	2.00	24	92	17	71	.81	1.31
August 1978 - January 1979															
St. Croix x St. Croix															
	6 months	9	9	100	8	89	14	1.56	1.75	14	100	11[e]	78	1.22	1.38
	1 yr and older	18	12	67	8	44	18	1.00	2.25	18	75	16[e]	75	.89	2.00
Rambouillet x Rambouillet															
	1.5 yr old	20	20	100	14	70	19	.95	1.36	18	90	15[e]	79	.75	1.07
Rambouillet x St. Croix															
	1 yr old	19	19	100	16	84	23	1.21	1.44	22	96	21[e]	95	1.10	1.31

[a] 40-day breeding periods beginning February 1 and August 1.   [b] Weaned at 60 days of age; percent of lambs born live.

[c] Breeding-lambing interval with 40-day breeding period.   [d] Age at beginning of breeding-lambing interval.

[e] Data based on preweaning age.

Source: R. C. Evans, K. E. Panter, A. J. Svejda, and W. C. Foote, Utah State University, unpublished data, 1979.

TABLE 4.2.4.   CONSECUTIVE LAMBING RESPONSE OF ST. CROIX EWES PLACED ON SIX-MONTH LAMBING INTERVALS IN UTAH

| Age when breeding was initiated | No. of ewes | Proportion of ewes lambing consecutively[a] | | | | | |
| | | January 1978[b] | | August 1978 | | January 1979 | |
		No. lambing	% lambing	No. consecutive lambings	% lambing	No. consecutive lambings	% lambing
6 months	6	6	100	2	33	0	0
1.0 year and older	10	10	100	8	80	5	50

[a] The overall mean lambing interval for all ewes was 233 ± 10.9 days with a range from 185 to 359.

The mean gestation length (34 ewes) was $149.9 \pm .58$ days.

The percentage of ewes bred when exposed to rams at 6-7 months of age (25 ewes) was 100%.

[b] Ewes had lambed at various periods of the year prior to this scheduled breeding period.

Source: R. C. Evans and W. C. Foote, Utah State University, unpublished data, 1979.

TABLE 4.2.5.   REPRODUCTION AND PRODUCTION PERFORMANCE BY BREED OR CROSS IN FLORIDA

Items	Breed or cross (1977-1978)			Breed or cross (1978-1979)		
	St. Croix	St. Croix x Florida Native	Florida Native	St. Croix	St. Croix x Florida Native	Florida Native
Number of ewes	9	45	45	13	49	50
% ewes lambing	89	87	91	77	94	90
No. lambs born	12	49	49	14	52	54
Lambs born/ewes bred	1.33	1.09	1.09	1.08	1.06	1.08
Lambs born/ewes lambing	1.50	1.26	1.20	1.40	1.13	1.20
Mean birth wt, kg	2.7	2.8	2.8	2.6	2.8	2.9
Lambs lost at 24 hours	1	1	2	4	0	0
Lambs lost at 60 days	1	5	5	4	1	1
Lambs lost at 120 days	1	5	5	4	2	2
Lamb Hb level	(67)[a] 10.3	(62) 10.2	(62) 10.4	(60) 9.3	(62) 10.2	(57) 9.7
Lamb Hb level	(118) 12.0	(113) 12.5	(113) 13.0	(118) 9.8	(120) 11.1	(115) 10.8
Lamb weight, kg	(64) 11.6	(59) 11.2	(59) 12.8	(62) 9.7	(64) 13.4	(59) 12.7
Lamb weight, kg	(118) 18.5	(113) 18.0	(113) 20.1	(117) 15.5	(119) 21.0	(114) 19.7
Mean lambing date	Dec. 22	Dec. 27	Dec. 27	Dec. 28	Dec. 26	Dec. 31

[a] Number in () indicates age in days when measurements were made.

Source: P. E. Loggins, University of Florida, unpublished data, 1979.

TABLE 4.2.6. LAMBING PERFORMANCE OF ST. CROIX AND BARBADOS EWES IN OHIO

Date exposed to rams	Ewes			Lambs			Lambing rate per:	
	No. exposed	No. lambing	% lambing	No. born	No. survived	% survived	ewe exposed	ewe lambing
St. Croix								
Older ewes								
9/76	10	10	100	17	15	88.2	1.70	1.70
8/77	8	8	100	17	16	94.1	2.12	2.12
2/78	8	2	25	3	3	100.0	0.38	1.50
7/79	8	8	100	16	16	100.0	2.00	2.00
Lambs								
8/77	7	2	29	2	1	50.0	0.28	1.00
Yearlings								
2/78	7	3	43	5	4	80.0	0.71	1.67
7/78	7	5	71	9	9	100.0	1.29	1.80
Barbados[a]								
Older ewes								
9/76	9	9	100	14	11	78.6	1.56	1.56
8/77	9	6	68	9	6	66.7	1.00	1.50
2/78	6	4	68	5	5	100.0	0.83	1.25
7/78	4	2	50	3	3	100.0	0.75	1.50
purchased,'78	4	2	50	4	4	100.0	1.00	2.00
Lambs								
8/77	9	0	0	0	0			
purchased,'78	3	1	33	1	1	100.0	0.33	1.00
Yearlings								
2/78	7	3	43	4	2	50.0	0.57	1.33
purchased,'78	4	4	100	7	6	85.7	1.75	1.75

[a] Ewes were variable in age and not necessarily typical of the breed.

Source: Charles F. Parker, Ohio Agricultural Research and Development Center, unpublished data, 1979.

Utah. Each of the 25 ewe lambs in the breeding flock was bred during the breeding period, indicating that 100% had reached puberty by 6 to 7 months of age.

The mean postpartum interval in the St. Croix breed during the fall was 31.6 days (SE = 5.18) based on data from 5 ewes. A high proportion of ewes allowed mating within 1 to 2 days following parturition. None of these matings was fertile and thus all were considered to be anovulatory, although the ovaries were not observed.

The length of breeding season has not been measured specifically in the United States because of management imposed to maximize numbers and to measure performance on 6-month lambing intervals. The St. Croix breed reportedly cycles throughout the year in the Virgin Islands. During the first winter in Utah, open ewes stopped cycling or failed to return to estrus at the expected time postpartum in March, which corresponded generally with the initiation of seasonal anestrus in wooled breeds. This finding indicates that the St. Croix and the wooled breeds have similar responses to latitude/photoperiod and climatic changes. The mean estrous cycle length was 16.21 (SE = 0.12) for all ewes tested.

Endocrine profiles (luteinizing hormone and progesterone) have been reported as similar to those of other breeds of sheep during the estrous cycle (Panter and Foote 1978; Bazer and Loggins, University of Florida, unpublished data 1979) and during late prepartum and early postpartum (Panter and Foote 1978).

Birth weights are shown in Tables 4.2.1 and 4.2.6. Overall mean birth weight ranged from 2.6 to 2.9 kg based on information from Utah and Florida.

Postnatal lamb loss varied among locations and was greatest in Utah. This might be due (at least partially) to the generally higher prolificacy recorded in Utah (Table 4.2.1 and 4.2.3).

Body weights at different ages and rates of gain are shown in Tables 4.2.5 and 4.2.7. These data indicate that the St. Croix sheep were lighter in weight than were the Florida Native sheep from birth to approximately 115 days. In comparisons made in Utah, the St. Croix sheep were markedly lighter in weight than were the Rambouillet at birth, as well as during a feeding period from approximately 80 to 122 days of age. St. Croix × Rambouillet crossbred lambs from Rambouillet ewes were comparable in weight to straightbred Rambouillet lambs. At the beginning of the feeding periods, weights for the 3 groups were 18.0, 30.5, and 27.0 kg, respectively, for the St. Croix, the St. Croix × Rambouillet cross, and the Rambouillet. At the end of the feeding period, the weights were 28.9, 42.8, and 41.9, respectively. Rates of gain measured at two-week intervals varied from 0.21 to 0.33 kg per day for the St. Croix, from 0.15 to 0.36 kg for the crossbreds, and from 0.32 to 0.34 kg for the Rambouillet.

Tables 4.2.8 and 4.2.9 summarize some carcass characteristics. The dress-

TABLE 4.2.7.   WEIGHT PER AGE AND RATE OF GAIN FOR ST. CROIX,
                RAMBOUILLET, AND ST. CROIX-RAMBOUILLET
                CROSSBRED RAM LAMBS

		Breed or cross	
	St. Croix	St. Croix x Rambouillet	Rambouillet
No. of lambs	13	12	7
Mean age at beginning of feeding period, days[a]	80	89	70
Weight at beginning of feeding period, kg[b]	18.0+.74	30.5+1.23	27.0+2.20
Weight at end of feeding period, kg	28.9±1.11	42.8± .94	41.9±2.64
Average daily gain over six-week feeding period, kg	.259	.292	.355

[a] Not adjusted for age.
[b] First body weights were taken when lambs were placed on feed.
    Ration consisted of pelleted alfalfa hay and rolled barley.
Source: R. C. Evans and W. C. Foote, Utah State University,
         unpublished data, 1979.

ing percentage for the older and younger age groups was 43.5 and 37.7%, respectively; this apparently reflects a reduced tendency for the St. Croix to deposit fat, even though the two groups of lambs were 287 and 194 days old. The amount of fat in the carcass was indicated by the covering over the medial area of the eye muscle (2.8 and 2.5 mm) and the maximum fat depth at the point of the ribs (11.2 and 4.3 mm) for the older and younger groups, respectively. There was a tendency, as expected, for the lambs to deposit fat as age increased. In Florida, comparisons were made with the Florida Native and the St. Croix × Florida Native cross. There were no major differences for the carcass traits studied.

In Florida, all sheep were managed without the use of anthelmintic medication. Hemoglobin levels were used as a measure of adaptation to this stress. The hemoglobin levels for ewes shown in Table 4.2.2 indicate no differences among ewes of the St. Croix, the Florida Native, or their crosses. Table 4.2.5 shows little or no difference among the groups of lambs for 1978–1979, but the groups may have differed during 1977–1978. This information suggests that the St. Croix may be able to produce well in environments with high, or relatively high, internal parasite loads.

SUMMARY

Twenty-two ewes and three rams of the St. Croix breed of sheep were brought to Utah State University, Logan, Utah, from St. Croix, U.S. Virgin Islands, in 1975. Several of the ewes were pregnant; thus the foundation flock was based on several sires. Samples of sheep from this flock were sent

TABLE 4.2.8.   CARCASS MEASUREMENTS ON ST. CROIX RAM LAMBS, MEAN ± SE

Age days	No. lambs	Live wt. kg	Chilled carcass wt. kg	Dressing %	Carcass grade[a]	Fat depth		Kidney fat kg	Leg wt. kg
						Eye muscle cover (medial) mm	Max. at point of rib mm		
287	4	45.5+1.34	19.27+.85	43.5+.05	G- to G+	2.8+.25	11.2+.51	.59+.06	5.43+.17
197	9	39.6+1.20	14.9+.83	37.7+.01	G- to Ch	2.3+.25	4.3+1.0	.59+.22	4.23+.18
Combined	13	40.77	16.08	39.4	G- to Ch	2.5	6.4	.43	4.60

[a] G- = low good, G+ = high good, Ch = choice.

Source: R. C. Evans and W. C. Foote, Utah State University, unpublished data, 1979.

TABLE 4.2.9.   YIELD GRADE MEASUREMENTS OF ST. CROIX, FLORIDA NATIVE, AND CROSSBRED LAMBS

Lambs	No.	Slaughter wt. kg	Carcass wt. kg	Dressing %	Fat[a] mm	Yield grade	Carcass length cm	Carcass depth[b] cm	REA[c] sq cm
1977 lambs[d]									
St. Croix	4	34.8	16.4	47.1	0.4	3.7	61.4	26.7	11.1
Native	4	33.7	15.7	46.6	0.5	3.6	60.0	26.0	9.9
1978 lambs[e]									
St. Croix	5	37.1	19.9	53.6		2.0	62.7	27.6	10.7
Native	6	37.4	18.0	48.5		2.1	60.3	28.1	12.6
Crossbred	7	37.9	19.8	52.2		2.3	62.4	28.4	12.3

a Fat over loin eye.

b Measured from top of shoulder at 5th rib to outside of breast.

c Rib eye area.

d Wethers.

e Rams.

Source: P. E. Loggins, University of Florida, unpublished data, 1979.

to Ohio and Florida in December 1976, and performances of the purebred St. Croix and certain crosses were compared with those of other breeds in all 3 locations.

The sizes of the sheep increased in the United States: in Utah, mature ewes averaged 54 kg and mature rams 74 kg. St. Croix ewe lambs showed high fertility at 6 to 7 months of age. Litter size for mature ewes varied from 1.5 at Florida (where the ewes were lighter and there was no treatment for internal parasites) to above 2.0 in the Utah flock. Lamb survival varied among locations but, in general, was comparable to that expected of temperate breeds of sheep raised in these environments. Of a small group of ewes one year old or older that were exposed to rams for 1 month, at 6-month intervals, 50% lambed 3 times in 18 months; no ewes first exposed to this mating scheme at 6 months of age lambed in all of the first three seasons. Mean lambing interval for all ewes was 233 days, with a range of 185 to 359 days.

St. Croix lambs weighed about 60% as much as did Rambouillet lambs. Rate of gain of lambs to 5 months of age was approximately proportional to birth weight for the two breeds.

Many of the studies conducted on the St. Croix also include other breeds with which they have been crossed or otherwise compared. These data provide an opportunity to compare some reproduction and production traits of the St. Croix with other breeds and also to observe some aspects of their combining ability in crosses.

REFERENCES

Panter, K. E. and W. C. Foote. 1978. Reproduction and endocrine function in the St. Croix ewe. *Proc. West. Sec. Am. Soc. Anim. Sci.* **29**:290.

THE BARBADOS BLACKBELLY ("BARBADO") BREED IN TEXAS

Maurice Shelton, *Texas A&M University Agricultural Research and Extension Center (San Angelo), U.S.A.*

The Barbados Blackbelly sheep in the United States are thought to have been introduced by the U.S. Department of Agriculture in 1904. These sheep quickly became concentrated in Texas, especially within the Edwards Plateau geographical region. Data on their numbers are not available since statistical reports do not show a classification by types. However, Texas probably has the largest collection of Barbados-type sheep in the United States. The author estimates that the number of sheep peaked at about 200,000 to 300,000 in the early part of the 1970s. In the late 1970s, this total was reduced markedly through slaughter and export to Mexico and other Central American and Caribbean countries. Three factors contributed to this reduction: (1) demand for hair sheep in the more tropical regions of Central America and the Caribbean, (2) generally increased demand and price for red meat of any source, which stimulated sale for slaughter, and (3) general revival of interest in wooled sheep in those areas where Barbados were raised. (Barbados sheep could not compete with more traditional breeds in the production of meat and fiber for the U.S. market, a point that will be discussed later in the text.)

Although the actual numbers of Barbados Blackbelly introduced into the United States are not known, it can be assumed that there were few and that the population has been increased by crossing with other breeds. A number of factors suggest that this crossing was primarily with Rambouillet or Merino types, which were the most prevalent types in the area of later Barbados concentrations. The original Barbados Blackbelly was apparently a polled animal, but in the United States it has been converted to a horned animal, with characteristics similar to those of finewool sheep. The lack of color variation suggests that Barbados Blackbelly sheep crossed extensively with Blackface (Suffolk and Hampshire) breeds. There is little evidence of continued crossing or mixing with finewool sheep, as the vestigial fleece cover that was characteristic of these animals in earlier years has been largely eliminated.

Several conditions may have contributed to the concentration of these sheep in the Edwards Plateau region. Perhaps they were first introduced as a novelty and then utilized as a meat supply for the home. They were often consumed by producers who usually did not choose to eat the meat of the more traditional breeds of sheep. This suggests a qualitative difference favoring the Barbados meat, probably associated with its relatively less fat and the tendency of the breed to store the fat internally. Since the unique tastes of red meats tend to be associated with the fat content, producers probably considered the taste of the Barbados meat similar to that of goat meat.

With the development of sport hunting and game farming as an industry in Texas, Barbados producers tended to shift their objectives toward that use. Most flocks have had some infusion of Mouflon breeding to provide a more suitable game animal. Although this breeding was done intentionally, and sometimes repeatedly, matings have been random in subsequent generations, and the Barbados remains the dominant influence. Apparently, the Barbados are more adaptable, more fertile, less seasonally restricted in breeding, and less subject to certain disease and parasite conditions than are the Mouflon. The relative concentration of Mouflon in the flock can be estimated by observations of color, temperament, and conformation. Sheep carrying significant Mouflon breeding will seldom have a black belly; ewes will tend to be fawn-colored, and males will often have white patches or saddles. Such sheep will have wilder temperaments and a more streamlined form.

As game animals, the male Barbados-like (mixed) sheep may be hunted on the ranch where produced. (They are sometimes called "wild Corsican rams.") For this purpose they have the advantage of not being classified as a native game animal, and thus have no seasonal hunting restriction. Another, and more widespread, practice is to gather the mature males and sell them to game farms or hunting clubs scattered throughout the country. For this market they usually sell at a price per head equal to or above that of domestic sheep sold for meat production. However, they must be kept to an older age than is usual for meat production, and there is a low harvest rate of huntable males. Most such flocks can be gathered for marketing, but with difficulty.

These sheep have not been utilized extensively in traditional research programs; thus there is a general lack of performance data. Mature ewes seldom reach 100 pounds, and lamb growth rates are slow.

Preliminary statements in this report suggested that the Barbados breed could not compete with more traditional breeds in the production of meat

TABLE 4.3.1.   SHRUNKEN LIVE WEIGHT AND LITTER SIZE FOR
BARBADOS EWES

Ewe age	Weight, kg		Litter size	
	No.	x̄	No. lambs in litter	No. ewes
Lamb teeth	10	21.9	1	19
Yearling	5	28.6	2	14
2 and 3 years	10	29.8	3	1
Solid mouth	7	31.3		
Aged	2	30.8	Mean = 1.47	

and fiber for the U.S. market. In the case of fiber, such statements are self-explanatory since these sheep are not shorn. In the case of meat production, the pure Barbados type will not grow sufficiently rapidly or produce a lamb carcass with the required conformation and fat cover to grade U.S. Choice. Thus, these sheep sell at a considerable disadvantage on the U.S. market, usually bringing less than one-half the price of U.S. Choice lamb. Other production parameters do not overcome this disadvantage. The ewes are not without some seasonal restriction to mating; they can be compared to Spanish or meat-type goats that have evolved for a number of years at a similar latitude, approximately 30° N. Although lambs may be born at any time of the year, most are born in winter or early spring. Some ewes that breed early in the season can be bred again so as to lamb twice in a 12-month period; however, one lambing per year is more common.

Table 4.3.1 shows preliminary data collected on a group of purchased ewes at the San Angelo Station in the winter of 1978–79. Weights were taken on arrival and would be lower than normal weights for these ewes.

4.4

CROSSBREEDING WITH THE "BARBADO" BREED FOR MARKET LAMB OR WOOL PRODUCTION IN THE UNITED STATES

Maurice Shelton, *Texas A&M University Agricultural Research and Extension Center (San Angelo), U.S.A.*

Although crossbreeding has been used in establishing the "Barbado" breed in the United States, it has not been used widely in routine production programs. Because some crossbreeding occurs accidentally, however, there is need to know what use can be made of the crossbred animals. Crossbreeding work has been done on an experimental basis by the Texas Agricultural Experiment Stations at McGregor and San Angelo. The F_1s involved were produced by crossing Barbados rams with Rambouillet ewes; therefore, they are similar to those produced by most accidental crossings. In commercial programs, such crossings might be produced more economically by reciprocal mating.

Data on the feedlot performance and carcass traits of F_1 Barbados × Rambouillet wether lambs have been reported by Shelton and Carpenter (1972). Straight Rambouillet rams, as well as Blackface (Hampshire, Suffolk, or Hampshire × Suffolk cross) rams bred to Rambouillet ewes were used for comparison. Table 4.4.1 summarizes the results.

These findings confirmed that Barbados-sired F_1 lambs can be graded as Choice market lambs for the U.S. market. However, their slower rate of growth means that they would be older upon reaching comparable weights. Assuming that they went into the feedlot at comparable weights (32 kg), the Barbados-sired lambs would require an additional 20 days of feeding. This conforms to feeders' observations and accounts for the usually lower selling price of these sheep. The Barbados-sired lambs generally are distinguishable by a hairy appearance, with a bare belly and bare legs (in some instances), and red or dark colors.

Because of their slower growth rate, these lambs might well require an ad-

TABLE 4.4.1. FEEDLOT PERFORMANCE AND CARCASS TRAITS OF BARBADOS-SIRED LAMBS COMPARED TO FINE WOOL OR CROSSBRED LAMBS

		Feedlot performance					Carcass traits			
Breed type	No. animals	Avg daily gain, g	Daily feed intake, g	Feed/gain	Dressing %	Wt. kg	Est. % consumer cuts	Rib-eye area, cm^2 /23 kg carcass	% kidney and pelvic fat
Rambouillet	14	251	1474	5.87	54.7	22.3	83.8	12.8	4.43
Blackface x Rambouillet	20	254	1397	5.51	56.8	23.5	82.9	12.6	3.57
Barbados x Rambouillet	20	176	1125	6.39	55.7	22.0	82.2	11.9	6.17

TABLE 4.4.2. PERFORMANCE OF DIFFERENT TYPES OF EWES EXPOSED
TO SUFFOLK RAMS CONTINUOUSLY FROM SEPTEMBER 1972
TO DECEMBER 1976 UNDER RANGE CONDITIONS

Breed or cross	No. of ewes	Mean annual fleece wt. kg	Mean annual body wt. kg	Mean no. lambings per ewe per year	Mean no. lambs born per ewe per lambing	No. lambs weaned per ewe per year
Rambouillet	25	4.0	51.0	1.12	1.41	1.22
Finnish Landrace x Rambouillet	38	2.7	52.8	0.98	1.80	1.27
Karakul x Rambouillet	24	2.8	55.3	1.19	1.34	1.33
Barbados x Rambouillet	24	1.7	46.8	1.24	1.71	1.64

ditional 2 to 3 months to reach a slaughter weight and condition comparable to that of other U.S. Choice lambs. The most notable difference in carcass traits was a higher percentage of kidney or pelvic fat in the Barbados-sired lambs. The Barbados rams used in these studies had not been selected for meat production; however, the other two breeds were highly selected. There seems little doubt that the Barbados breed would respond to selection for growth and carcass traits.

The ewe offspring produced within the previous study were placed in an accelerated lambing study along with straight Rambouillet, Finnish Landrace × Rambouillet, and Karakul × Rambouillet lambs, with the results reported by Thompson and Shelton (1978). The ewes and rams grazed together continuously under arid range conditions with a supplemental feeding program designed to keep the ewes in strong condition. This management system was not successful, as shown by results in Table 4.4.2.

These data suggest that Barbados × Rambouillet ewes adapt better to accelerated lambing programs than do the other types tested under these conditions. They lambed more frequently than did the other types and had a lambing rate that approached that of the Rambouillet × Landrace ewes. The wool produced by both the Karakul-cross and Barbado-cross ewes had relatively lower value in the traditional market channels. Both fleeces were a mixture of coarse hair and wool fibers, with some off-colored fibers. These types of wools might have unique value in the homespinning trade but would require special efforts in merchandising (Gallagher and Shelton, 1977). The lambs of these two types would probably not be accepted readily by buyers.

The lambs produced by the Karakul-cross ewes had a noticeable fat accumulation around the tail. Lambs from Barbados-cross ewes bred to blackface rams had growth rates lower than those of lambs from Rambouillet ewes. The lambs from the Barbados-cross ewes also were colored atypically.

If a realistic value is placed on the weight of wool and of the early-weaned lambs produced, the income realized from both the Karakul- and the Barbados-cross ewes would slightly exceed that from Rambouillet ewes. However, in neither case would this value compensate for the lowered value of the F_1 ram or wether lambs, which are a by-product of producing the F_1 ewes. Thus, in traditional ranching programs in the Edwards Plateau, these types do not appear to have advantages.

REFERENCES

Gallagher, J. R. and M. Shelton. 1977. Fiber traits of primitive and improved sheep and goats. Texas Agric. Expt. Sta. PR-3190.

Shelton, M. and Z. L. Carpenter. 1972. Feedlot performance and carcass characteristics of slaughter lambs sired by exotic breeds of rams. Texas Agric. Expt. Sta. PR-3024.

Thompson, P. V. and M. Shelton. 1978. Accelerating lambing potential in Rambouillet and exotic type crossbred ewes. Texas Agric. Expt. Sta. PR-3446.

BARBADOS BLACKBELLY SHEEP IN MISSISSIPPI

Leroy H. Boyd, *Mississippi State University, U.S.A.*

Barbados sheep were first introduced into Mississippi during the early 1960s when 16 sheep of breeding age were purchased from a ranch near Kerrville, Texas, and moved to a ranch near Winona in north-central Mississippi. Within a decade the flock numbered over 500 head. Most of the flock then was sold, with the exception of 49 ewes and 16 rams of mixed ages that were delivered to Mississippi State University in the spring of 1971. The body weight of these mature ewes ranged from 34 to 43 kg and that of the mature rams from 60 to 68 kg. All ewes were polled and all rams were horned, except for one ram lamb born in the project. All original stock had the black-belly color pattern.

Mississippi State University is located at 33° 38′ N and 88° 48′ W, at an elevation of 85 m. Average maximum temperature in August is 33° C and average minimum temperature in January is 1° C, whereas rainfall averages 1,320 mm per year. Mean relative humidity is 75% during the hotter months (June to August).

During early April 1971, a group of Dorset ewes was placed with the Barbados ewes and three Barbados rams. Although they were penned together at night, the two breeds would separate while grazing.

No Barbados-sired lambs were produced from the Dorset ewes. After a Dorset ram was placed with the flock, only straightbred Dorset and Barbados lambs were produced. Mixed-breed mating and conception occurred only after the introduction to the breeding flock of Dorset and Barbados ram lambs that had been reared together.

Other observations on the Barbados: Barbados lambs appeared more susceptible to enterotoxemia than Dorset lambs; newborn Barbados lambs would approach humans but their mothers ran away; Barbados ewes generally lambed during the day without assistance.

COMPARISON OF BARBADOS, DORSET, AND
CROSSBRED EWE PERFORMANCE

The adaptability and usefulness of the Barbados, Dorset, and Barbados ×
Dorset ewes were studied in a multiple-lambing program in Mississippi. Average mature weights for producing ewes were: Barbados, 45.3
kg; Dorset, 63.5 kg; and crossbred, 56.7 kg. For the 44 ewe lambs born
during the winter of 1972–73, age at first breeding period (October 15 to
December 1, 1973) ranged from 8 to 10 months. Other breeding periods
were April 15 to June 1 and August 1 to September 5. All dry and lactating
ewes were exposed to rams during the fall breeding period; only dry ewes
were exposed during the other breeding periods. Two or three mature Suffolk rams weighing 113 to 136 kg were with the ewes during each breeding
period.

Dorset and Barbados × Dorset ewes were shorn in late March and
again in July each year. Average yearly wool production was 3.6 kg for
Dorset and 3.2 kg for the crossbred ewes. Crossbred wool was coarse and
varied widely in quality, with some shedding occurring on the belly and
neck. The Barbados ewes were never shorn and the little wool that they produced was shed each spring.

Ewes were drenched before breeding, just after lambing, at weaning, after
the summer shearing, and at two other times during the year. At birth,
lambs were weighed and permanently identified. All lambs were docked.
Ram lambs were castrated when 1 or 2 weeks old.

A combination of ryegrass, oats, and subterranean clover pasture was
grazed from October 15 to June 1, except in 1977 when grazing was ter-
minated on December 15, 1976, and resumed on March 1, 1977. Grain was
fed to ewes after they lambed. Equal amounts of shelled corn and a 20%
crude protein, high-energy, commercially available pelleted feed were fed at
a level of 0.45 kg per head daily during lactation. Summer pastures consisted primarily of common bermuda and dallisgrass. A creep feeder containing the same pelleted feed was available for the lambs. After grazing
about 7 hours each day, all sheep were put in shelters during the night.

Table 4.5.1 shows the ewe reproductive performance by season. Ranked
on the basis of number of ewes lambing during the five possible lambing
periods, the Barbados were most dependable, followed by the Barbados × Dorset cross and Dorset ewes. Table 4.5.2 gives added data about the
ewes.

By the age of 4 years, Barbados and Barbados-cross ewes had each
averaged approximately 4.4 lambings, compared to only 3.1 lambings for
the Dorset. The litter size of the Barbados ewes was substantially larger
than that of the Dorset, with the litter size of crossbred ewes almost exactly

TABLE 4.5.1. REPRODUCTIVE TRAITS FOR BARBADOS, DORSET, AND BARBADOS-DORSET CROSSES IN DIFFERENT SEASONS (1974-77)

| | Lambing period[a] | | | | | | | | | |
	Spring 1974	Fall 1974	Winter 1974-75	Spring 1975	Fall 1975	Winter 1975-76	Spring 1976	Fall 1976	Winter 1976-77	Total
Barbados[b]										
Ewes lambing	15	-	15	-	13	1	9	2	11	66
Lambs born	22	-	19	-	24	2	20	3	22	112
Lambs weaned	22	-	19	-	24	2	17	3	22	109
Barbados x Dorset[a]										
Ewes lambing	11	1	12	2	11	3	8	5	8	61
Lambs born	12	1	13	2	18	4	15	10	15	90
Lambs weaned	11	1	13	2	16	4	15	10	14	86
Dorset[a]										
Ewes lambing	8	5	6	5	4	8	1	8	2	47
Lambs born	8	5	6	7	6	13	2	9	4	60
Lambs weaned	7	5	5	6	4	13	2	6	4	52

[a] Correspond to breeding periods of October 15 - December 1, April 15 - June 1, August 1 - September 15 each year.
[b] See Table 4.5.2 for number of ewes in each breed group.

TABLE 4.5.2. REPRODUCTIVE PERFORMANCE OF BARBADOS,
              BARBADOS X DORSET, AND DORSET EWES EXPOSED
              TO SUFFOLK RAMS DURING THREE PERIODS PER
              YEAR FOR THREE YEARS

Trait	Barbados	Barbados x Dorset	Dorset
No. ewes			
Started on exp.	15	14	15
Died during exp.	2	1	2
No. lambings			
Potential[a]	70	69	71
Actual/potential, %	94	88	66
Avg litter size			
Born	1.70	1.48	1.28
Weaned, 60 days	1.65	1.41	1.11

[a] All 44 ewes potentially could have lambed in the spring of 1974; subsequently, however, only dry ewes were exposed to rams during the
breeding season and thus might potentially lamb in the next period.

TABLE 4.5.3. TYPE OF LITTER AND WEIGHTS OF SUFFOLK-SIRED
              LAMBS FROM BARBADOS, DORSET, AND CROSSBRED
              EWES (1974-77)

	Barbados				Barbados x Dorset				Dorset			
	Male		Female		Male		Female		Male		Female	
Litter type	No.	Wt	No.	Wt	No.	Wt	No.	Wt	No.	Wt	No.	Wt
Birth wt, kg												
Singles	10	4.1	12	3.9	11	4.9	22	4.4	21	4.3	14	3.7
Twins	44	3.2	40	3.2	32	3.9	22	3.5	19	3.4	6	3.1
Triplets	4	2.3	2	2.3	2	3.6	1	2.9	-	-	-	-
60-day wt, kg												
Singles	10	18.2	12	17.2	11	22.0	21	19.2	15	22.0	13	20.4
Twins	44	13.6	40	13.6	29	18.4	22	16.3	18	17.1	6	18.5
Triplets	2	13.4	1	15.4	2	11.8	1	10.0	-	-	-	-

intermediate. The lamb survival percentage of the Barbados and the crossbred ewes was higher than that of the Dorset ewes, in spite of the Barbados
ewes' larger litter size. The number of lambs weaned per 4-year-old ewe was
3.5, 6.1, and 7.3 for the D, B × D, and B ewes, respectively.

Average individual lamb weights at 60 days (computed as the unweighted
mean of male and female singles and twins) were 19.5, 19.0, and 15.6 kg for
D, B × D and B groups, respectively (Table 4.5.3). Data on the lambs from
the B × D ewes indicate approximately 8% heterosis for 60-day lamb weight

TABLE 4.5.4.   PERFORMANCE AND CARCASS TRAITS OF
SUFFOLK-SIRED LAMBS FROM BARBADOS,
BARBADOS X DORSET, AND DORSET DAMS

Item	Ewe breed or cross		
	Barbados	B x D	Dorset
No. of lambs	18	16	10
Slaughter age, days[a]	220	226	200
Slaughter wt, kg[a]	42.3	46.3	42.3
Carcass wt, kg	22.4	23.9	21.5
Dressing percent	53.1	51.7	50.9
Carcass wt per day of age, g	102	106	108
Quality grade[b]	12.9	13.8	13.4
Leg score[b]	12.4	13.2	13.2
Loin eye area, sq. cm	13.5	13.7	13.8
Backfat thickness, mm	6.1	5.6	4.8
Hindsaddle, %	45.6	45.8	46.8
Kidney/pelvic fat, %[a]	5.5	4.1	2.9
Yield grade[a,c]	3.9	3.6	3.0

[a] Differences among dam groups were significant (P<0.05).
[b] 11 = Choice, 14 = Prime.
[c] Lower yield grade score = higher estimated lean yield.

as a maternal trait. Heterosis for birth weight as a maternal trait was approximately 15%.

Total 60-day weight of lambs per lambing was 21.6, 25.8, and 23.8 kg for the D, B × D, and B ewes, respectively, with the corresponding values per potential lambing being 14.3, 22.8, and 22.4 kg. Thus, the crossbred ewes equaled the straight Barbados ewes when both numbers and weights of lambs were considered.

Lambs born in March-April, 1974, were fed and slaughtered to obtain carcass data. After weaning at 60 days of age, these lambs grazed common bermuda grass and regal clover, and were fed a daily ration of 0.45 kg of the pelleted feed per head. They were sheared and drenched at weaning and again in early August; the lambs then were sorted by breed of dam and placed on full feed. The free-choice ration consisted of the pelleted feed and mature timothy–red clover hay. Lambs were slaughtered at the MSU Research Meats Facility on a weekly basis from late September to November 1, 1974.

Table 4.5.4 shows the performance and carcass traits for lambs born in March-April, 1974. Differences between wethers and ewes were not significant; therefore these data were combined. Lambs from Barbados and Barbados-cross ewes were slaughtered at older ages than lambs from Dorset ewes. Thus, breed differences in other traits such as carcass weight, yield, and fatness may be due to the differences in slaughter age. There were significant breed differences (P < 0.05) for yield grade and (P < 0.01) for percent kidney-pelvic (KP) fat. The degree of Barbados breeding influenced yield grade and percent KP fat, with lambs from Barbados dams containing more of the KP fat that would contribute to a less desirable yield grade. Since lambs averaged High Choice to Low Prime in grade, Barbados lambs should maintain the Choice grade if slaughtered at younger ages to obtain leaner carcasses with more desirable yield grades.

SUMMARY

Barbados and Barbados × Dorset ewes had more frequent lambings and larger litter sizes than did Dorset ewes, when given the opportunity to lamb 3 times in 2 years. Lamb survival of Suffolk crossbred lambs also was higher for the B and B × D ewe groups. Crossbred ewes showed heterosis for fertility, lamb birth weight, and lamb weaning weight, but not for litter size. Total weight of lambs at 60 days per ewe exposed was about equal for the B and B × D groups; these total weights were more than 50% heavier than those from D ewes. Lambs from B and B × D ewes were fatter than those from D ewes at similar weights, but the two groups with Barbados breeding were older and had achieved a higher percentage of mature weight when slaughtered. If ewe reproduction, lamb growth, and carcass characteristics are taken into account, the results suggest that the Barbados × Dorset ewes are the most productive of the three genotypes compared for an accelerated-lambing, terminal-cross program. The numbers in the experiment were small, but the average of 6.4 lambs weaned per crossbred ewe in a 3-and-a-half-year period suggests that such animals have good potential for intensive lamb production in the southern United States.

BARBADOS BLACKBELLY SHEEP IN CALIFORNIA

J. M. Levine and G. M. Spurlock,
University of California, Davis, U.S.A.

A small flock of Barbados Blackbelly sheep was formed in 1971 on the G. M. Spurlock ranch at Dixon, California. The sheep were purchased in Texas and privately maintained by Spurlock. Dixon is located at 38° N and is one of the northernmost locations for which data have been reported on the Barbados Blackbelly breed.

This study investigated whether the Barbados ewe retains its ability to lamb throughout the year at nearly 40° N. The sheep were raised in drylot, primarily on alfalfa hay. First-parturition ewes bearing more than one lamb were allowed to raise only singles, with the remaining lambs raised artificially. For all subsequent lambings, ewes were allowed to raise singles or twins, with remaining lambs raised artificially. Lambing interval was used as a selection criterion to produce a flock of ewes that would lamb throughout the year.

Phenotypically, the sheep were mostly of blackbelly pattern with some black, and some yellow (tan) or light coloring. A few had patches of wool.

For the period March 1971 to September 1977, data were reported on age at first lambing, lambing intervals, litter size, gestation period, and lambing by season. A total of 275 lambings by 64 ewes was recorded over the 6.5 year period.

RESULTS

Age at First Lambing

Average age at first lambing was 370.4 days for 64 ewe lambs. Most ewes lambed for the first time between 300 and 420 days of age with a range from

TABLE 4.6.1. LITTER SIZE FOR FIRST AND LATER PARTURITIONS

Parturition	No. of each litter size				Total ewes lambing	Mean litter size	SD
	1	2	3	4			
First	38	25	1	0	64	1.42	.53
Second and later	46	132	29	1	208	1.93	.61
All	84	157	30	1	272	1.81	.63

229 to 513 days. Thus, ewe lambs began to cycle at 4.5 to 5 months of age and were bred by 9 months of age. Only 3 ewes lambed for the first time at less than 300 days old, and 9 ewes did not lamb until more than 420 days old.

Litter Size

Table 4.6.1 summarizes the data on litter size. Average litter size for first lambing was 1.42. Selected ewes had an average of 1.93 lambs/ewe for second and all subsequent lambings; the litter size increase from first to later lambings can be attributed to increased maturity of the ewes and to selection. For all lambings, litter size averaged 1.81. Of the first lambings, 59% of the litters consisted of single lambs, 39% twins, and 2% triplets. Of all subsequent lambings, 22% of the litters were singles, 63.5% twins, 14% triplets, and 0.5% quadruplets. For mature ewes (second and later lambings), almost 78% of the lambings were multiple births. Thus, under California conditions and with selection, the Barbados Blackbelly sheep were shown to be very prolific.

Relationship Between Litter Size at Lambing (n)
and Litter Size at Next Lambing (n + 1)

Table 4.6.2 summarizes the relationship between litter size for ewes at lambing (n) and litter size at their next lambing (n + 1). Only data for ewes bearing two successive litters are included for these comparisons. Effect of parity is not considered. Ewes that bore singles at their nth lambing averaged 1.62 lambs for their next (n + 1) lambing; ewes that bore twins for lambing (n) had an average of 1.92 lambs at next lambing; and ewes that bore triplets for lambing (n) averaged 2.25 lambs at next lambing. Ewes that bore triplets averaged 0.3 more lambs in the next litter than did ewes that bore twins (2.25 vs 1.92). Ewes that bore twins also averaged 0.3 more lambs per litter at next lambing than did ewes that bore singles (1.92 vs 1.62). These results suggest that ewes that were relatively more prolific for early litters tended to remain so. However, ewes were well fed at all times and no ewe was al-

TABLE 4.6.2. RELATIONSHIP OF LITTER SIZE FOR ONE LAMBING, N, AND LITTER SIZE AT NEXT LAMBING, N+1

Litter size for lambing n+1	Litter size for lambing n			
	1	2	3	4
1	23[a]	19	3	0
2	30	95	12	0
3	2	9	9	1
Total lambings (n+1)	55	123	24	1
Mean litter size	1.62	1.92	2.25	3.00
SD, litter size	.56	.51	.68	–

[a] No. of ewes bearing singles that bore singles at next lambing.

TABLE 4.6.3. CORRELATIONS AMONG LITTER SIZES

Litter sequence	No. litters	Litter sequence		
		1	2	3
2	53[a]	.31		
3	41	.26	.33	
4	33	.06	.03	.18

[a] No. of ewes with two or more lambings.

lowed to suckle more than two lambs, factors which likely contributed to the consistent prolificacy.

Correlations Among Litter Sizes

All correlations among litter sizes (Table 4.6.3) were positive and ranged from 0.33 (between second and third litters) to 0.03 (between second and fourth litters). Correlation coefficients for adjacent litters tended to be higher than those for nonadjacent litters. This suggests that lactation stress had little effect on ewe fertility. As indicated in Table 4.6.2, litter size was positively related to litter size at next lambing.

Lambing Interval

Table 4.6.4 summarizes data for lambing interval. Fifty-four ewes had an average of 222.6 days between lambings, or slightly more than 7 months. Thirty-five percent of the intervals were between 181 and 200 days in length, 22.5% between 201 and 220 days, 12.5% between 221 and 240 days,

TABLE 4.6.4. LAMBING INTERVAL FOR EWES EXPOSED
CONTINUOUSLY TO RAMS UNDER DRYLOT
CONDITIONS AT LATITUDE 38° N

	Ewes with 2 or 3 litters	Ewes with 4 or more litters	All ewes
No. ewes	21	33	54
No. intervals	29	171	200
Mean, days	232.4	221.0	222.6
SD, days	48.7	41.1	42.4
Range, days			172–378

TABLE 4.6.5. EFFECT OF LITTER SIZE ON SUBSEQUENT LAMBING
INTERVAL

	Litter size			
	1	2	3	4
No.	55	121	23	1
Mean interval, days	219.2	221.1	238.4	261
SD, days	38.7	43.1	47.0	–
Range, days	177–355	173–369	182–378	–

9.5% between 241 and 260 days, and 8.5% between 261 and 280 days. Only 12% of the lambing intervals fell outside the range of 181 to 280 days (3.5% were less than 181 days and 8.5% were greater than 280 days). Twenty-one ewes that remained in the flock and had fewer than 4 litters had an average lambing interval of 232.4 days, while 33 ewes that had more than 4 litters had an average lambing interval of 221 days. This suggests that selection for decreased lambing interval in the flock had some effect and that a flock of this breed with lambing intervals of 7 months can be developed fairly readily under northern California drylot conditions.

Effect of Litter Size on Subsequent Lambing Interval

Table 4.6.5 shows little difference in subsequent average lambing interval for ewes with single lambs or with twins (219.2 vs 221.1 days), but lambing interval was greater for ewes that had triplets (238.4 days). Apparently, under these environmental conditions, Barbados ewes can bear singles or twins with little effect on subsequent lambing interval; however, ewes that bear triplets, even though nursing only two lambs, would have longer intervals.

TABLE 4.6.6. EFFECT OF PARITY ON LAMBING INTERVAL

Interval between parities	No. ewes	Lambing interval, days Mean	SD
1-2	54	241.7	45.7
2-3	39	204.9	32.1
3-4	30	220.9	41.4
4-5	26	229.1	53.7
5-6	20	206.0	25.3
6-7	17	217.9	32.1
7-8	8	216.6	23.5
8-9	5	225.4	57.4
9-10	1	201.0	-

TABLE 4.6.7. EFFECT OF LITTER SIZE ON GESTATION PERIOD, DAYS

	Mean litter size 1	2	3	All
No.	28	50	5	83
Mean gestation	151.0	150.2	150.2	150.5
SD	2.6	1.6	1.8	2.0
Range	146-158	143-153	148-152	143-158

Lambing Interval by Litter Sequence

Table 4.6.6 shows data for average lambing interval between successive parities. The longest interval (241.7 days) occurred between first and second lambings, and the shortest (discounting the record of the single ewe that had 10 lambings) was between second and third lambings. Differences among other lambing intervals were small and could be explained by sampling error. These data indicate both a high genetic potential and a favorable environment for frequent lamb production because high prolificacy and a short lambing interval were maintained for up to 10 lambings.

Gestation Period

Table 4.6.7 shows data for gestation period by litter size. Average gestation length was 150.5 days, differing little for litters of 1, 2, or 3 lambs.

Lambing by Season

Of 262 lambings by 64 ewes between March 1971 and September 1977, 23% occurred during the spring, 19% during summer, 19% during autumn, and 40% during winter (Table 4.6.8). Thus, after 6.5 years at 38° N, Barbados ewes were lambing throughout the year (although with greatest frequency during winter and spring).

TABLE 4.6.8. DISTRIBUTION OF LAMBINGS BY SEASON FOR THE
PERIOD MARCH 1971 TO SEPTEMBER 1977 (64 EWES)

	Spring 3/21–6/20	Summer 6/21–9/22	Autumn 9/23–12/21	Winter 12/22–3/20
No. lambings	60	49	49	104
%	23	19	19	40

TABLE 4.6.9. LITTER SIZE BY SEASON

	Spring	Summer	Autumn	Winter
		First lambing		
No. lambings	11	4	2	28
Mean litter size	1.24	1.33	2.00	1.53
SD	.44	.58	–	.56
		Subsequent lambings		
No. lambings	49	45	46	78
Mean litter size	2.05	1.81	1.79	2.11
SD	.72	.61	.46	.68

Litter Size by Season

Table 4.6.9 summarizes data for litter size by season. Most first litters were
born during winter and spring. For mature ewes, there was a correlation
between seasonal incidence of lambing and litter size. Lambings were most
frequent in winter, with the three other seasons about equal in frequency of
lambings. Litter size also was highest during winter, with a mean of 2.11.
This finding suggests that the Barbados breed in California is not completely aseasonal; rather it exhibits higher reproduction performance during
late summer/early autumn breeding seasons as measured by increased
lambings and litter size in winter/spring. This performance could be due to
changes in feed, to climate, or to a photoperiod response; however, the
causes are not known.

Annual Lamb Production Per Ewe

Annual lamb production per ewe was 2.96 over 6.5 years with an average
lambing interval of 222.6 days and an average litter size of 1.81. Calculated
on the basis of mature ewes only (with average litter size of 1.93), lamb production per ewe per year was 3.16 lambs. This is higher than for any
temperate zone breed, with the possible exception of the Finnish Landrace.
The fact that this number of lambs is born in 1.6 litters rather than the single

litter per year should result in a substantially higher number raised per year by the Barbados ewes. The data from this study indicate that Barbados ewes are capable of producing at this level (78% multiple births every 7 months) over an extended period of time.

SUMMARY

Data were available from 275 lambings by 64 Barbados ewes maintained in drylot on a private ranch at Dixon, California (38° N) between March 1971 and September 1977. Average age at first lambing was 370.4 days. Litter size was 1.42 for first lambings, 1.93 for subsequent lambings, and 1.81 for all lambings. Ewes that were more prolific at one lambing tended to be more prolific in subsequent lambings. Lambing interval averaged 222.6 days. Ewes that bore singles had a shorter average interval to next lambing (219.2 days) than did ewes with twins (221.1) or triplets (238.4). Gestation period averaged 150.5 days and was not influenced by size of litter. Of 262 lambings, 40% were in winter, 23% during spring, and 19% each during summer and autumn.

Litter size tended to be higher in winter and spring. For second and later lambings, litter size was 2.11 for winter, 2.05 for spring, and 1.8 for both summer and fall. Calculated on the basis of overall average litter size (1.81) and lambing interval (222.6 days), Barbados ewes produced 2.96 lambs per ewe per year. Mature ewes averaged 3.16 lambs per ewe per year, with an average litter size of 1.93.

INDEX

Africana sheep, 17
African sheep
 origin of Colombian, 79
 prolificacy of, 83
 reproduction in, 81
 size of, 83(table)
 See also Hair sheep; West African
  hair sheep
Agricultural Development Corporation
 (Jamaica), 177
American hair sheep
 differences among, 14–15
 West African origin of, 13–14
Animal protein, 202
Appalachian Mountain Research Station,
 270–274

Bahia, Brazil, sheep productivity in,
 130(table)
Balami sheep
 carcass characteristics of, 216(table)
 reproductive performance of, 215(table)
Barbados, sheep distribution from,
 193–195
Barbados Blackbelly crosses
 accelerated lambing study of, 270–274
 carcass traits of, 303(table)
 ewe performance of, 300–304
 kidney fat of, 296
 lamb weights of, 302(table 4.5.3)
 parasite resistance of, 264–270
 performance of, 263(table), 271,
  295(table), 303(table)
 prolificacy of, 302(table 4.5.3)
 reproductive performance of, 260(table
  4.1.2), 302(table 4.5.2)
 reproductive traits of, 301(table)

Barbados Blackbelly sheep, 15–16
 African ancestor of, 180–182
 birth weights of, 122(table), 157(table),
  168–169, 169(table 2.10.5)
 breed characteristics of, 152
 breeding results, 157–159
 in California, 305
 carcass traits of, 194, 291, 294(table),
  303(table)
 crossbreeding of, 259–264, 289,
  293–296
 development of, 189–192
 estrus synchronization of, 109(table)
 ewe characteristics of, 158(table 2.9.4)
 ewe performance of, 300–304
 ewe weights of, 145, 146(table 2.8.2)
 feedlot performance of, 294(table)
 fertility of, 161(table 2.9.8)
 first lambing age of, 305–306
 heat stress resistance of, 264
 in Jamaica, 177–178
 lamb birth weights of, 110(table)
 lambing distribution for, 159(table)
 lambing interval for, 166–168,
  167(figure), 168(table), 307–308,
  308(table 4.6.4), 309(table 4.6.6)
 lambing performance of, 178(table
  2.12.1), 282(table)
 lamb mortality, 41, 116(table 2.5.14),
  169(table 2.10.6)
 lamb weights for, 112(table),
  146(table 2.8.1), 302(table 4.5.3)
 litter size, 37, 108(table), 121(table),
  147(table), 159(table), 160(table
  2.9.6), 164(table), 164–166, 166(table
  2.10.3), 306–307, 307(table 4.6.2),
  310(table 4.6.9)
 management of, 300

mating plan for, 120
in Mississippi, 299
Mouflon breeding, 290
origins of, 152, 179–180
performance of, 178(table 2.12.2),
  303(table)
perinatal mortality of, 68–70
productivity advantage of, 47
prolificacy of, 68(table), 161(table
  2.9.8), 192, 302(table 4.5.3)
reproductive performance of, 302(table
  4.5.2)
reproductive traits of, 301(table)
in Tobago, 142
in United States, 30, 289
weaning weight of, 158(table 2.9.3)
wooled ancestor of, 182–187
and wooled breed crosses, 160–161
and wooliness, 192
Barbados Sheep Station, 152–154
Bergamasca breed, 128
Birth weight, 39(table 1.2.8), 41
  comparisons of, 110–111
Blackhead Persian sheep, 18–19, 82
  in Barbados, 191
  ewe weights, 145, 146(table 2.8.2)
  lamb mortality rates, 116(table 2.5.14)
  lamb weights, 110(table), 112(table),
    146(table 2.8.1)
  litter size of, 108(table), 147(table)
  size of, 83(table)
  in Trinidad, 142
Blenheim Sheep Station (Tobago),
  144–145
  sheep performance at, 145–148
Body weights, 41–42
Booroola Merino, litter size of, 37
Bouaké Research Center (Ivory Coast),
  228
Brazil
  hair sheep breeds in, 18, 125–127,
    133–137
  production environment of, 127–128
  sheep population in, 125, 126(table)
Brazilian Somali Blackhead sheep, 127
  reproductive traits of, 132(table 2.7.5),
    133(table)
  weights of, 136(table 2.7.13)
Breeding management, 120
  in Barbados, 154–156
Breeds, 7–19

Cameroons, Forest hair sheep in, 212
Climate, and estrus, 98(figure)
Coat characteristics, 4–7
Colombia, 17, 18, 79–84
Creole. *See* Mixed breed sheep
Criollo sheep, 85–86
  estrus synchronization, 109(table)
  lamb birth weights of, 110(table)
  lamb mortality rates of, 116(table
    2.5.14)
  lamb weaning weights, 112(table)
  litter size for, 108(table)
  in Venezuela, 106
Crossbreeding programs, in Barbados,
  156–157

Disease, 25
Djallonké sheep, 212
  birth weight of, 235(table 3.3.7)
  description of, 227
  fecundity of, 232
  gestation period for, 231(table 3.3.3)
  lamb growth rate for, 236–237,
    236(table 3.3.9)
  lambing distribution of, 230(table)
  lambing interval for, 231–232
  lamb mortality in, 233(table),
    237–238, 238(table)
  lamb performance, 234–238
  management practices for, 228–229
  performance traits of, 242(table)
  prolificacy of, 232
  reproductive performance of, 229–234
  reproductive traits of, 233(table)
  weight and body measurements of,
    228(table)
  See also Forest hair sheep
Dorset sheep, 161(table 2.9.8)
Drax, James, 186

Ebini Station (Guyana), 119
Efficiency, 31–32
Environment, 29
Estrus synchronization, tests on, 109–110
Ewe weights, 44(table)
Exotic sheep, 156

Farmers, 50
Feed production, 50
Fertility, 31
  and nutrition, 33

postpartum-period effect on, 64(table 2.1.8)
Finnish Landrace sheep, 259–264
Fitness, 28–30
Flemish sheep, 187
Flock Efficiency Index (FEI)
  defined, 46
  rankings for, 47
Flock productivity, 45–49
Flock Productivity Index (FPI)
  rankings for, 47
  traits in, 46
Forage intake, 221
Forest hair sheep, 211–212
  characteristics of, 11(table)
  compared with Nungua Blackhead sheep, 211(table)
  of West Africa, 7–13

Gaitana Farm, La, 82–84
Genetic improvement, 50
  adaptation/production tradeoff, 25
Genotypes, interaction with environment, 29
Gestation length, 32, 33(table)
  and litter size, 64(table 2.1.9)
Goats, 3
Golden Grove Estate (Barbados), 163–164
Grazing management, 153
Growth rate, comparisons of, 114

Hair fibers, 4
Hair sheep
  adaptability of, 85
  coat types, 6
  compared with wool sheep, 19–21
  crossbreeding of, 6
  ewe flock management, 107(table)
  heterogeneity of, 20
  local names for, 10(table)
  performance measures in, 108–116
  phenotypic variability, 82–83
  tropical adaptation of, 29
  uses of, 3–4
  See also names of specific breeds
Heat stress, 258
Heterotypes. *See* Hair fibers

Ivory Coast, 227

Jamaica, 177–178, 187–188

Katahdin breed, 16
Kemp fibers, 4
Kemp sheep, 187
Kolda Station (Senegal), 241

Lambing interval, 32–34, 34(table)
Lamb mortality, 37–41, 38(table)
  and litter size, 39(table 1.2.7)
  in Tobago, 148(table)
Larceny, 143
Ligon, Richard, 180, 184–186
Litter size, 34–37, 35(table)
  distribution among breeds, 36(table)
  repeatability of, 165

Maine, 16
Mali (central), 245–248
Management practices
  in Barbados, 154–156
  traditional, 26
  in Virgin Islands, 172
Market demand, for sheep meat, 26
Marketing
  in Brazil, 129–130
  infrastructure, 50–51
  in Tobago, 144
Mexico, 14, 17, 55–73, 188
Milk yield, 38–40
Mixed breed sheep, litter size of, 121(table)
Mococha Livestock Experiment Station (Mexico), 57–58
Morada Nova sheep, 18, 125–126
  lamb performance of, 136(table 2.7.11)
  lamb weights, 135(table), 136(table 2.7.12)
  litter size, 37
  prolificacy of, 131
  reproductive traits of, 131(table), 132(table 2.7.5)
  weights of, 134(table), 137(table)
Mortality rates, comparisons of, 115–116

Netherlands, 187
Nigeria
  climate in, 201, 219
  hair sheep in, 219–220
Nkolbisson Station (Cameroons), 212
North Carolina, 257
Nungua Blackhead sheep, 211(table)

Palmar Farm, El, 80–82
Parasites, 25
  in Barbados, 155–156
  resistance to, 264–270
Pelibuey sheep, 17
  birth weights of, 77(table)
  characteristics of, 56–57
  colors of, 14
  at first lambing, 62–63, 63(table)
  flushing effect on prolificacy, 67(table 2.1.12)
  grazing productivity of, 59(table)
  growth characteristics of, 70–72
  lambing interval for, 67(table 2.1.10), 78(table 2.2.5)
  lamb mortality, 41
  management/feeding of, 58
  mating schedule for, 60(table)
  Mexican population of, 55
  milk yield of, 78(table 2.2.6)
  perinatal mortality, 68–70
  prolificacy of, 68(table)
  at puberty, 61, 62(table)
  reproduction in, 58–60, 61–70, 75–78
  reproductive performance of, 76(table)
  size characteristics of, 56(table)
  See also Hair sheep
Peliguey sheep. *See* Pelibuey sheep
Pelo do Boi, 3, 18
Performance traits, summary of, 32–45
Permer sheep, crossbreeding of, 209–211
Peul sheep, body weight of, 242(table)
Production systems
  constraints on, 24–26, 128
  livestock-based, 24
  mixed crop/livestock, 23–24
  in northeast Brazil, 128
  in Tobago, 143
Production traits, 28–32
  in Brazil, 130–131
  breed averages for, 48(table)
  in tropics, 29(table)
Productivity, 26–28
  and constraint removal, 27(figure)

Rabo Largo sheep, 127
Red African hair sheep, 86
Reproduction efficiency, energy-level effect on, 64(table 2.1.7)
Research and development, 49–51
Ruminants, 21(n1)

Sahel hair sheep
  birth weights of, 253(table)
  characteristics of, 11(table)
  description of, 249–250
  flock structures for, 250(table 3.5.2)
  growth of, 253(figure)
  lambing characteristics of, 252(table)
  ownership pattern, 250(table)
  reproductive performance of, 251–254
  of West Africa, 7–13
  See also Hair sheep; West African hair sheep
St. Croix sheep
  body weights of, 277(table), 278(table)
  carcass measurements on, 285(table)
  fertility of, 276–278
  hemoglobin levels for, 278(table)
  lambing performance of, 282(table)
  lambing response of, 280(table)
  reproductive characteristics of, 279(table)
  reproductive performance of, 281(table)
  in United States, 275
  weight of, 284(table)
  yield grade measurements of, 286(table)
St. Elizabeth breed, 177
Santa Inês sheep, 18, 126–127
  Bergamasca origins of, 133
  lamb weights of, 136(table 2.7.12)
  prolificacy of, 131
  reproductive traits of, 131(table), 132(table 2.7.5)
Savanna hair sheep
  characteristics of, 11(table)
  of West Africa, 7–13
  See also West African Dwarf sheep
Senegal, 241–243
Sexual maturity, by age/weight, 43(table)
Sheep
  in Americas, 187–189
  domesticated, 3
  ownership, 151
Shika Station, Kaduna State (Nigeria), 213
SODEPALM farm, 239–240
Suffolk sheep, 161(table 2.9.8)

Tobago, 141–142
Tobago Government Farm, 142
Touabire sheep, body weight of, 242(table)

Tsetse flies, 201

Uda sheep, 212–214
  body weight of, 213(table 3.1.14)
  carcass characteristics of, 216(table)
  ewe performance of, 213(table 3.1.13)
  reproductive performance of,
    214(table), 215(table)
University of Ghana Experiment Station
    (Nungua), 211
University of Ibadan, 202–203
University of Ife farm, 220

Venezuela, 17, 18, 85–104, 105–117
Virgin Islands (U.S.), 171–172
Virgin Islands White hair sheep, 16
  body weights/measurement of,
    174(table)
  characteristics of, 173
  litter size, 175(table)
  origin of, 172–173
  performance of, 173–175

Weaning weight, 41–42, 42(table), 155
  comparisons of, 111–112
West Africa, ecozones in, 9(figure)
West African Dwarf sheep, 7, 9, 202
  birth weights of, 205(table 3.1.6),
    207(table)
  breeding performance of, 222–224
  carcass traits of, 208–209, 210(table),
    216(table)
  crossbreeding with Permer sheep,
    209–211
  diseases of, 224
  ewe productivity, 203–206
  feeding/management of, 220–222
  first lambing age of, 203(table)
  lambing interval of, 204(table 3.1.3)
  lamb performance, 206–208
  lamb weights, 146(table 2.8.1)
  milk yields of, 208(table 3.1.8)
  mortality of, 206–208
  parasites of, 224
  prolificacy of, 205(table 3.1.4),

    205(table 3.1.5)
  reproductive performance of, 215(table)
  reproductive traits of, 204(table 3.1.2)
  retail cuts for, 209(table)
  weights of, 208(table 3.1.9), 221(table)
  See also Djallonké sheep; Savanna
    hair sheep; West African hair sheep
West African hair sheep
  age/weight at puberty, 91(table 2.4.4)
  birth weight of, 101(table)
  in Caribbean, 18
  estrous cycle, 90(table 2.4.2), 91, 109
  ewe weights of, 146(table 2.8.2)
  horned, 14
  lamb birth weights of, 110(table)
  lamb mortality in, 97–99, 99(table),
    100(table 2.4.14), 116(table 2.5.14)
  lamb weaning weights, 112(table)
  litter size in, 96(table), 108(table),
    147(table)
  pasture lamb growth of, 102(table)
  performance of, 87–88
  reproduction of, 87, 88–97, 93(table)
  seasonal variation in lambing, 97(table)
  and season effect on reproduction,
    94(table)
  size of, 89(table)
  Venezuelan origin of, 86
  See also Hair sheep
White Fulani sheep. *See* Yankasa sheep
Wild sheep, 4–6
Wiltshire Horn sheep, 191
Wool fibers, 4
Wool sheep
  coat types, 6
  compared with hair sheep, 19–21
  crosses with hair sheep, 6
  tropical adaptation of, 30

Yankasa sheep, 212–214
  body weight of, 213(table 3.1.14)
  carcass characteristics, 216(table)
  ewe performance of, 213(table 3.1.13)
  reproductive performance of,
    214(table), 215(table)

About the Book and Editors

Hair Sheep of Western Africa and the Americas:
A Genetic Resource for the Tropics
edited by H. A. Fitzhugh and G. E. Bradford

Perhaps less than 10 percent of the world's sheep have hair coats instead of wool fleeces, but as an animal resource, these sheep are important far out of proportion to their numbers. Hair coats provide an advantage in the face of the heat, humidity, and other stresses of the tropics. Small in body size, early to mature, and often highly prolific, hair sheep can well serve the needs of small-scale farmers; they also have potential for larger-scale commercial production to supply meat for urban markets.

The list of contributing authors is an excellent guide to hair sheep experts currently active in field research in the developing nations. Based on their firsthand experience, the reports in this application-oriented collection examine the various strains of hair sheep, their fertility and growth rates, and other performance characteristics. The editors add an overview of the place of hair sheep in the world today, summarizing and comparing performance statistics for different breeds and strains and reviewing breeding and management options for hair sheep producers.

H. A. Fitzhugh is an animal scientist and program officer for Latin America and the Caribbean at Winrock International Livestock Research and Training Center. **G. E. Bradford** is a professor of animal science at the University of California, Davis.

Printed and bound by CPI Group (UK) Ltd, Croydon, CR0 4YY

23/10/2024

01778263-0005